P9-CQL-232

The Political Economy of Oil

The Political Economy of Oil

Ferdinand E. Banks
The University of Uppsala,
Sweden
Monash University,
Melbourne, Australia

LexingtonBooks
D.C. Heath and Company
Lexington, Massachusetts
Toronto

24890

Library of Congress Cataloging in Publication Data

Banks, Ferdinand E
The political economy of oil.

Bibliography: p.
Includes index.
1. Petroleum industry and trade. I. Title.
HD9560.5.B263 333.2'7282 79-3340
ISBN 0-669-03402-9

Copyright © 1980 by D.C. Heath and Company

All rights reserved. No part of this publication may be reproduced or transmitted in any form or by any means, electronic or mechanical, including photocopy, recording, or any information storage or retrieval system, without permission in writing from the publisher.

Published simultaneously in Canada

Printed in the United States of America

International Standard Book Number: 0-669-03402-9

Library of Congress Catalog Card Number: 79-3340

For Madeleine Louise Banks

Contents

List of Figures

List of Tables

Preface and Acknowledgments

This text and reference book on the economics of oil contains a large amount of material I have used in my courses in resource economics and economic theory and in my various lectures on oil and associated issues. Particular attention is paid to the many questions I have been asked on this topic, some of which convinced this economist that anyone who thinks that he or she can comprehend the seriousness of the oil crisis without touching on such topics as the international financial system and the effect of the oil-price rises on investment and productivity is the victim of a delusion, and a serious one at that. Chapter 8 summarizes this book and, to a certain extent, my approach to the subject.

I want to make it clear that energy is one of the three most important issues in economics today. The other two, which will probably be even more important in the long run, are the economics of nonfuel minerals and of population. Our civilization will stand or fall on the basis of what the decisionmakers in the industrial countries learn about these topics in the next few decades. Accordingly, I can only hope that this book helps wean at least a few potential decisionmakers away from the overembellished trivia filling the most learned of the learned journals and leads them down the path of sweet relevancy.

I wrote this book to be of use to a broad range of readers, and so, with one or two exceptions, technical material is confined to chapter appendixes. Still, I must confess that foremost in my mind were the interests of the reader who has completed a course or two in economics. Let me emphasize, though, that at least half of this book can be read profitably by people who are interested in but have never formally studied economics.

As usual, I have had a great deal of help with this project from colleagues, critics, and friends: Lennart Berg (who kept me from repeating a particularly embarrassing mistake), Bengt Hansson, Carl Gustav Melén, and Yngve Andersson of the University of Uppsala; Lars Bergman of the Stockholm School of Economics; and Marian Radetzki of the Institute for International Economics, Stockholm. I have benefited greatly from the comments of Professors Mike Folie of Shell Australia; Ron Jones of the University of Rochester, who held the first Gunnar Myrdal Professorship at the Institute for International Economics; and Bruno Fritsch of the Federal Institute of Technology, Zurich. I received an excellent introduction to energy forecasts for the long term from Dr. Wolf Häfele, who is in charge of research at the Institute for Systems Analysis, Vienna. I benefited greatly from a lecture given by Professor Robert Pindyck in Stockholm early in

1980 and particularly from being able to examine Olle Björk's important survey of the world oil market. I would also like to thank the Reserve Bank of Australia once more for their Professorial Fellowship, because while I was in Sydney I first became fully exposed to some of the topics in this book. I am equally grateful to the participants of seminars at Uppsala Institute of Management, Berlin. I could not possibly have completed this book without travel grants from Uplands Banks of Uppsala and the Swedish Social Science Council (Samhällsvetenskapliga Forskningsräd).

Finally, I thank the Centre for Policy Studies, Monash University, Melbourne, and its director, Professor Michael G. Porter, for giving me an opportunity to continue under their auspices, my work with energy and resources. I would also like to direct some attention to Jo L., who almost, but not quite, got "his" book into circulation.

1 Introduction

This is a text and reference book. Its specific purpose is to discuss, at an elementary level, the political economy of the world oil market. This means not only petroleum economics per se but also, to a limited extent, various aspects of macroeconomics and international monetary economics which are essential to understanding the oil crises, past, present, and future.

However, despite the impression that might be created by simply thumbing through the pages, this book is nontechnical. Abstract techniques are strictly confined to chapter appendixes and to one or two sections (indicated by asterisks) that contain slightly technical but elementary presentations. Here I can say that much of the material in chapter 6 has been used in a course in international economics that I occasionally give beginning students with only a few months' background in economics. They had no more difficulty in understanding it than did my graduate students in mathematical economics and in many cases less difficulty, because they were unconcerned about relating this material to those elegant but empty models that still characterize a large part of the graduate economics curriculum. The reader will find a summary of the book in chapter 8.

In terms of language and layout, what follows is designed so that the reader can gain a maximum amount of knowledge about the economics of oil (and cognate phenomena) with a minimal effort. Still, some effort is necessary. For instance, the reserve/production ratio is a pivotal issue in petroleum economics, and perhaps a pencil, paper, hand-held calculator, knowledge of grammar school arithmetic, and a small amount of concentration would be useful in examining those pages in chapters 3 and 4 that treat this topic. Similar attentiveness is also required at various other points in the text, although the presentation is completely straightforward. Finally, I would like to take this opportunity to affirm the power and importance of elementary economic theory as it is presented in most textbooks and in this book, despite its astonishing failure to help clarify some of the economic dilemmas that have confronted the world since 1973, and despite the fact that academic economics is gradually becoming a kind of algebraic circus that everyone is tired of watching, but cannot leave for fear of offending the ringmasters.

Units and Equivalents

This section is deliberately brief to avoid boring the reader who wants to get to the heart of the subject with a minimum of delay. But even so, the reader who is not ready to look at a few numbers should simply skip it and proceed to the next section or, for that matter, go directly to chapter 2, which begins with Professor Friedman's famous misunderstanding about what was going to happen on the world oil market. But for those who are ready for a few numbers, we can begin by emphasizing that there are a variety of ways to present data about crude oil (or *crude petroleum* as it is more correctly called), and it is often convenient for the reader to be able to go from one set of units to another. The most popular unit for measuring both consumption and production is millions of barrels per day. For example, at present the United States and Western Europe consume about 19.5 and 15.5 million barrels per day (Mbbl/d), respectively. These figures can be turned into another popular unit, millions of tons per year, by multiplying by 50. Note also that these are *flows*: they contain a time dimension. But *reserves*, or the amount of recoverable oil in the ground at a point in time, are *stocks*. These are usually measured in billions of barrels or billions of tonnes. For instance, the most recent census of crude-oil resources, as of January 1976, is shown in table 1-1.

Next, we can ask just what is the relation between a tonne (or, metric ton) and a barrel. In a sense there is none, since a barrel has to do with volume and a metric ton is a unit of mass; but obviously if the barrel is filled with a liquid (such as oil), it has a weight. Determining this weight poses a

Table 1-1
World Crude-Oil Reserves
(billions of metric tons)

	Cumulative Production to 1976	*Known Reserves*	*Believed Reserves*[a]	*Total*	*Percentage of Total*
United States	16	7	11	18	7.0
Other Western Hemisphere	8	12	25	37	14.5
Russia, China, and other centrally planned	8	14	50	64	25.0
Middle East	12	68	19	87	34.0
Other Eastern Hemisphere	4	14	36	50	19.5
Total	48	115	141	256	100.0

Source: U.S. Geological Survey; The Petroleum Economist (Various Issues); The international Petroleum Encyclopedia, 1978.

[a]These are undiscovered reserves or reserves which geologists expect will be found.

problem, since the weight (and quality) of a liquid varies with its specific gravity and viscosity (which roughly measures its "fluidness"). Oil is rated according to an American Petroleum Institute (API) number or degree; the higher the API, the lighter the oil. Saudi Arabian Light oil has an API of 34.7. The weight-volume transformation used most often is 1 tonne = 1 metric ton (t) = 7.33 barrels (bbl). This is an average selected from a range that runs from 1 t = 6.98 bbl to 1 t = 7.73 bbl. Thus in table 1-1 the 7 billion metric tons of proved reserves held by the United States is equal to $7 \times 7.33 = 51.3$ billion bbl (or 51.3 Gbbl, where G is the prefix "giga-," meaning 10^9).

It should also be made clear that 1 t is the designation of a metric ton, which equals 2,204 pounds (lb). We also have a short ton, which in some countries is simply called a ton; 1 short ton (ton) = 2,000 lb. Thus 1 t = 1.1023 tons. Finally there is a long ton, which is 2,240 lb.

Equivalencies

As we know, oil is only one source of energy. Other sources are gas, coal, nuclear fission, solar energy, and so on. Therefore we need some means of putting these on an equivalent basis. For example, how much coal is equal to 1 cubic meter (m^3) of gas or 1 bbl of oil? This turns out to be a simple matter since all energy sources have an energy content which is quantifiable in terms of the amount of heat that the source generates. These sources include fossil fuels and fuel wood, which generate heat during combustion, and nuclear fuels, which generate heat by nuclear fusion in electric power plants. Heat is usually measured in British thermal units (Btu), where one British thermal unit is the amount of heat needed to increase the temperature of one pound of water by one degree Fahrenheit. For example, 1 t of anthracite coal has an energy content of 28,000,000 Btu, while 1 t of crude oil has an energy content of 42,514,000 Btu; and the energy content per ton of crude oil is nearly nine times that of unprocessed oil shale or tar sands. (Another convement unit is the *therm*, which is 100,000 Btu.)

Electric energy in its various forms can also be converted to British thermal units. As an example, the fuel equivalent of hydroelectricity (in 1970) was 10,500 Btu per kilowatt hour (Btu/kWh). With this as a background, it is interesting to examine table 1-2, which discusses expected energy demand for the Organization for Economic Cooperation and Development (OECD) countries in 1980 as well as forecast demand for 1985 and 1990. It has been argued that these forecasts, which originate with the OECD secretariat, suffer certain deficiencies; but as far as I can tell, they are representative of the forecasts of oil supply and demand made during the last few years and fairly useful when the limitations of this kind of exercise are understood.

Table 1-2
OECD Energy Balance for 1980, 1985, and 1990
(*million barrels of oil per day equivalent*)

	1980	*1985*	*1990*
OECD Availability			
Indigenous: Oil	16.8	14.5	14.5
Natural gas	14.9	15.0	15.0
Coal	15.9	16.5	21.0
Nuclear	4.2	6.0	9.0
Renewable	5.8	6.5	7.5
Synthetic fuel	0	0	1.5
Net imports of gas, coal, and other nonoil	0.5	4.5	—
Total	58.1	62.5	68.5
Total OECD Energy Demand	86.2[a]	92.0	103.0
Required oil imports into OECD	28.1	29.5	34.5
The imports of non-OPEC LDCs	2.4	2.0	2.0
Communist net imports	– 1.6[b]	1.0	—
OPEC consumption	2.6	4.5	5.4
Total demand for OPEC oil	31.5	37.0	41.9
OPEC supply	30.5	33.0	36.0
Shortfall (demand minus supply)	1	4.0	5.9

Source: OECD. *Energy Balances of OECD Countries*, Paris 1978; and reports of the IEA, 1978, 1979.
[a]This figure is probably too high because of a recession in the United States and parts of Western Europe in 1980.
[b]Minus sign signifies exports.

At present an attempt is underway to employ the International System of units (SI) to measure energy equivalents. Here the basic energy unit is the joule (J), although for electric energy the watthour (Wh) is allowable. Before we show equivalents, note that the basic multiple units are the gigajoule (GJ), petajoule (PJ), kilowatthour (kWh), and terawatthour (TWh). Thus

$$1 \text{ GJ} = 1{,}000{,}000{,}000 \text{ J} = 10^9 \text{ J}$$

$$1 \text{ PJ} = 1{,}000{,}000 \text{ GJ} = 10^{15} \text{ J}$$

$$1 \text{ kWh} = 1{,}000 \text{ Wh} = 10^3 \text{ Wh}$$

$$1 \text{ TWh} = 1{,}000{,}000{,}000{,}000 \text{ Wh} = 10^{12} \text{ Wh}$$

Note that the principal designations for sizes are as follows:

kilo	k	10^3	(thousand)
mega	M	10^6	(million)
giga	G	10^9	(billion)
tera	T	10^{12}	(trillion)
peta	P	10^{15}	(thousand-trillion)
exa	E	10^{18}	(million-trillion)

For example, 1 MWh = 10^6 Wh = 1,000,000 Wh and 1 TJ = 10^{12} J = 1,000 GJ.

Now it is possible to present a simple table for equivalents. Note that we usually measure the heating value of various fuels in terms of the heating value of oil, calling such and such an amount of coal, gas, or uranium equivalent to so and so many *t*ons of *o*il *e*quivalent (toe) or, in the majority of cases, millions of tons of oil equivalent (Mtoe = 1,000,000 toe). Thus we have

	PJ	*TWh*	*Mtoe*
Petajoules	1	0.2778	0.023885
Terawatthours	3.6	1	0.085985
Million tons of oil equivalent	41,868	11,630	1

It is also worth remembering that 1 Mtoe can be converted to British thermal units, should this unit be relevant to the discussion. A handy transformation is 1 bbl oil = 5,800,000 Btu and, as pointed out above, 7.33 bbl = 1 t. Also note that we are dealing with three basic forms of energy: electric (*e*), thermal (*th* or *therm*), and chemical (*ch* or *chem*). For example, 50 GWh_e = 50 GWh of electric energy.

The thing to understand now is that forecasts of the type reviewed in table 1-2 are based on very imperfect information. At present, for instance, it is being said that almost all forecasts of the oil that will be supplied by Saudi Arabia and Iran are too high. It also appears that nuclear supplies *could* make a sizable jump from their present low level if all the construction planned is completed on time; but, as is pointed out later, nobody really knows whether this will happen. In fact, another well-known organization, the Central Intelligence Agency (CIA), claims that an even slower rate of growth of energy supplies will take place over the first part of the 1980s than that shown in table 1-2. If this turns out to be true, then there will be very little scope for increased energy consumption in the major industrial countries.

On the demand side, the usual technique is to assume one or more rates of economic growth and associate energy use to these assumptions through some kind of relationship which postulates that for every percentage point of economic growth, a certain percentage growth in energy will be experienced. The usual ratio, or *elasticity,* is between 0.8 and 1.1. Thus until recently it was thought that OECD economic growth should attain a minimum value of 3.5 percent per year during the 1980s; if this were so, then the required growth rate of energy inputs (unless some draconian savings measures were introduced) should be about 3 percent per year. But given what *could* happen to supply, it seems realistic to expect some serious shortfalls (or gaps between demand and supply) at various times during the 1980s. Of course, shortfall is an a priori concept. Unless major new sources of petroleum are discovered or the largest producing countries decide to lift more oil, these gaps will be closed by income (and economic growth) declines that reduce demand, and/or major energy price rises that have the same effect and lead to a depletion of the extremely high inventories of oil now being held all over the world. Above all, it is doubtful whether price rises, by themselves, will cause most of the OPEC countries to raise their production. The reasons for this last contention are given later.

Refining (or Further Processing), and Sea Transportation

We conclude this short chapter by saying a few words about refining. A refinery is an installation that turns crude oil into a "slate" of well-known products. The most important are gasoline, kerosene, distillate fuel, and residual fuels. Refineries are extremely large economic units. It has been estimated that a refinery must be able to handle at least 100,000 to 150,000 barrels of oil per day (bbl/d) before it realizes any significant scale economies. In addition, as compared with the business of pumping oil, significant amounts of physical capital are required per barrel of oil processed. In the Western Hemisphere, the investment cost per barrel for a typical refinery seems to be increasing at a rate faster than the general rate of inflation, and some of the estimates of the amount of money required for both new and replacement investment in refineries in the United States seem to verge on the astronomical. At the same time it should be appreciated that the oil industry is not starved for profits and ranks near the top of the industrial league in terms of such things as rate of return on equity and rate of return on total capital. Just now, the average price of refining a barrel of crude is about $.75.

There is another important economic activity between the extraction and refining of a large part of the world's supply of crude oil.—sea transportation. Before 1973 sea trade in general grew at an average rate of 8 percent per year in volume and 11 percent in ton-miles. This discrepancy was largely the result of a faster-than-average growth in the shipping of crude oil. Oil is the paramount sea cargo, and oil tankers account for 50 percent of the world's shipping fleet, 40 percent of the total tonnage, and over 50 percent of annual ton-miles. The amount of oil transported by sea increased from 250 Mtons in 1954 to almost 2 Gtons in 1979. Over this period the world's tanker fleet increased from 3,500 ships to 7,000 and in total weight from 37 million deadweight tons (37 Mdwt) to 340 Mdwt. Despite an existing tanker surplus which came to 17 Mdwt at the end of 1979, tankers are still being built. For the most part these are smaller tankers (of 60,000 to 100,000 dwt) of which only a few are idle. One of the reasons is the deepening of the Suez Canal, which will mean that tankers can take the short route to Europe from the Arabian Gulf (thirty-one days) rather than the long route around the Cape (forty-one days). It was this long route that made large tankers (150,000 dwt and up) so economical when the Suez Canal was closed. Viewed over the last few years, the average cost of shipping a barrel of oil between the Middle East and Rotterdam has been about $1. These rates vary considerably, however. In the closing months of 1979, large tankers were getting freight rates of $1.22/bbl to $1.36/bbl, with their breakeven cost being close to the top of this range; in the beginning of 1979, they could get only $.137/bbl.

Two more observations are necessary. Although we know OPEC has been greatly concerned with raising its revenues without increasing the amount of oil it produces, at present there seems to be very little future in attempting to raise these revenues by increasing the price of oil, since the price increases of 1979 (which came to almost 100 percent) are the principal reason that most of the industrial countries, including the United States, are experiencing record levels of unemployment and anticipate that living standards generally are going to come under a severe downward pressure. However, OPEC could conceivably increase its cash flow by a large increment by becoming more heavily involved in the transporting and refining of oil. In a limited sense this may already be taking place, since it was recently announced that several of the larger oil companies were planning to cooperate with Saudi Arabia in major refining projects; and Kuwait has just announced that, where possible, the purchasers of its oil must transport this oil on ships owned by Kuwait.

Finally, freight rates are generally given in *worldscale flat rates*. These are established semi-annually by the International Tanker Nominal Freight Scale Association, and are the basic reference listing of tanker rates. The rate is given for each route, and is the approximate cost of shipping a barrel of crude oil over the particular route on a tanker of "standard size," which happens to be 19,500 long tons. Actual rates are then expressed as a percentage of the worldscale rate, and depend on the supply of and demand for shipping. For example, in February 1978 a 50,000 deadweight ton tanker carrying crude from the Persian Gulf to Japan quoted a "spot" (single trip charter) rate of worldscale (WS) 80. Thus the actual rate was $0.80 \times 1.26 = 1.008$ dollars/barrel, where 1.26 is the listed or WS rate.

Appendix 1A: Energy-Economy Interaction

The elementary flow model of basic economic theory has been used by Manne, Richels, and Weyant (1979) to discuss various aspects of energy forecasting. There is some algebraic confusion in their work, and it deals with long-run situations which, in the present world with its succession of oil price shocks, are much less interesting than "impact" effects, but they do raise several interesting issues. If we take $D_{\hat{e}} = f(p, X)$, where $D_{\hat{e}}$ is the demand for energy, p is the price of energy, and X is some aggregate variable such as gross national product, then the usual manipulations give us

$$\hat{D}_e = n_i \hat{X} + n_p \hat{p} \tag{1.1}$$

or

percent change in energy demand =

$$\begin{pmatrix}\text{GNP elasticity}\\ = n_i\end{pmatrix}\begin{pmatrix}\text{percent change}\\ \text{in GNP}\end{pmatrix} + \begin{pmatrix}\text{price}\\ \text{elasticity} = n_p\end{pmatrix}\begin{pmatrix}\text{percent change in}\\ \text{energy price}\end{pmatrix}$$

In these expressions n_i and n_p are commonly known as elasticities. Now if we take $n_i = 1$ and $n_p = -0.3$ (which seem to be the most widely used values), then it has been suggested that if the United States plans to maintain a long-term GNP growth rate of 3 percent, but intends to reduce energy growth to 2 percent, then energy prices would have to rise at a rate of $[(3 \times 1.0) - 2]/0.3 = 3.3$ percent per year in real terms.

How can output rise when the energy price is rising fairly rapidly and energy inputs are decreasing? The answer is that there is supposed to be a substitution of other factors for energy. It should be emphasized, though, that we are dealing with long-run phenomena which have important ramifications for wages, salaries, and real income, as is made clear in chapter 2. The reader should also realize that with present inflation rates a *real* price rise of 3.3 percent corresponds to an increase in *money* or *nominal* prices of 12 to 15 percent.

Next we must say something about the supply side. If we have an upward-sloping supply curve of the type $S_e = g(p)$, with $\mathrm{d}S/\mathrm{d}p^{6}0$, then we can write $\hat{S}_e = n_S\hat{p}$, where n_s is the elasticity of supply. Since we have $\hat{S}_e = \hat{D}_e$ when we have an equilibrium, we get $n_S\hat{p} = n_i\hat{X} + n_p\hat{p}$, or

$$\hat{p} = \frac{n_i\hat{X}}{n_s - n_p} \tag{1.2}$$

The estimates I have of n_S center on $+0.2$, and thus with $\hat{X} = 3$ percent, $\hat{p} = 1 \cdot 3/[0.2 - (-0.3)] = 6$ percent. (It should also be pointed out that if we look at the supply curve in isolation and ask the implications of a 2 percent growth in energy and the 3.3 percent increase in price we obtained earlier, we see that it means an elasticity of supply of $n_s = \hat{S}_e/\hat{p} = 2/3.3 = 0.606$.)

Now we have a $\hat{p}$ of 3.3 percent and another of 6 percent. There is no point in choosing between them since it is obvious that the price of energy can be restrained by shifts to the right of the supply curve. If we write $\hat{S}_e = h(p, \alpha)$, where α is a shift factor, we get $\hat{S}_e = n_S\hat{p} + n_\alpha\hat{\alpha}$. Long-run equilibrium means $\hat{S}_e = \hat{D}_e$, or

$$\hat{p} = \frac{n_i\hat{X}}{n_s - n_p} - \frac{n_\alpha\hat{\alpha}}{n_s - n_p} \tag{1.3}$$

Next we need a value for n_α; and to estimate this, I used some information from the 1960s when the price of energy in the United States was almost constant; energy growth 4.4 percent/year and GNP growth 2.75 percent. From equation 1.3, with $\hat{p} = 0$, $n_\alpha = n_i\hat{X}/\hat{\alpha} = 1 \cdot 2.75/4.4 = 0.625$. Using this value for n_α, we find that

$$\hat{p} = \frac{1 \cdot 3}{0.5} - \frac{0.625}{0.5} \cdot 2 = 6 - 2.5 = 3.5 \text{ percent}$$

Notice that in this calculation $\hat{\alpha}$ is approximated by 2 percent since this is the predetermined rate of energy growth regardless of what happens to the price of energy. Once again, the reader should be reminded that a *real* rate of growth of the price of energy that is equal to 3.5 percent implies a double-digit value for the rate of growth of this price in nominal or money terms. Some doubt exists in my mind as to whether a long-term GNP growth rate of 3 percent (in real terms) is possible under these circumstances.

Oil and the World Economy

On March 4, 1974, a distinguished Nobel Laureate, Professor Milton Friedman, was able to provide the readers of *Newsweek* with the joyful tidings that " . . . the world oil crisis is now past its peak. The initial quadrupling of the price of crude oil after the Arabs cut output was a temporary response that has been working its own cure." Furthermore, he was able to inform his audience that " . . . even if they [OPEC] cut their output to zero, they could not for long keep the world price of crude at 10 dollars a barrel. Well before that point the cartel would collapse." And, presumably, the price of oil would come tumbling down.

Today the average posted price of crude oil (or *crude petroleum,* which is the correct technical designation) is above $30/bbl outside the United States; the physical output of petroleum seems to be stagnating in many oil-producing countries; and commercial artists throughout the industrial world are at work designing functional and aesthetically pleasing coupons for the oil and fuel rationing that seems certain to be necessary, eventually. Friedman's assurances notwithstanding, the Organization of Petroleum-Exporting Countries (OPEC) still has not collapsed. (However, now it should be obvious that the growing internal strains in OPEC did not and could not lead to a fall in oil prices. On the contrary, these dissentions have meant a higher average price than otherwise would have been the case.) As far as I am concerned, it is the economies of the industrial world that are facing, and will continue to face, the prospect of an extended sojourn in the emergency ward, although in terms of such things as unemployment, some of the most important industrial countries have yet to recover from the 1973-1974 oil-price shock. Certainly, if the definition of the standard of living could be broadened to include demoralized expectations and lost options, then it appears likely that tens of millions of people in these countries are in a worse position today than they were five years ago.

One of the main purposes of this chapter is to attempt to convince the reader, in the simplest possible terms, that the oil crisis has propelled the world into an economic crisis and that regardless of occasional upturns, this impasse will continue as long as the industrial countries are susceptible to oil-price rises of the magnitude experienced during 1973-1974 and 1979. If we take the case of the United States, oil prices and imports have grown in such a way that the additional cost of imported oil resulting from a recent 35 percent oil-price increase was at least as much as the increased costs of oil imports resulting from the 400 percent price rises of 1973-1974. This helped contribute to a large depression in the profits of many U.S. businesses, since the decelerating U.S. economy made it difficult for heavy energy users to pass on these expenses. Because profits are a major source of funds for financing investments in plant and equipment, we can conclude that the spiraling oil price will inevitably have a deleterious effect on employment and/or wages in *all* sectors of the economy, as well as on employee morale: Increased investment and growth lead to a greater dynamism in a society because new opportunities are created in the form of new types of jobs (while at the same time some less desirable occupations are eliminated); and by making visible the rewards for study and work, the increased investment and growth tend to sharpen the motivation of many jobholders. It may also be true that if the price of oil is completely decontrolled, as people such as Friedman advocate, the ensuing increase in various household, transportation, and energy expenditures will lead to a fall in the real standard of living of many people in the United States, with the less affluent suffering most because this group spends a larger fraction of its income on items with energy-sensitive prices than the well-to-do. However, even the rich, particularly those in a hurry, should also experience some discomforts, since it appears that such modes of luxury travel as the Concorde may not be able to continue in service if fuel costs continue to escalate at their present rate.

There is also a very strong possibility that not many decades will pass before an analogous torment surfaces in the form of a tight supply of nonfuel minerals. This issue is treated in detail in Banks (1977, 1979b), but the following observation can be offered at present. Since energy is an increasingly important factor in the extraction and processing of various ores, it could happen that a sustained energy shortage (initiated by a downturn in the world production of oil) would result in a pronounced scarcity of some industrial raw materials or, more likely, an inflation in their prices of such a degree as to give the impression of physical exhaustion.

In these introductory remarks there has been a great deal of speculation about how soaring energy prices might undermine the material basis of industrial society. It thus seems reasonable that many economists broaching this possibility are asking themselves, and perhaps others, about the rest of the drama—the sequence where these price rises induce their own cure in the form of increased supplies and reduced consumption. There is no shortage of economic expertise ready to provide ironclad assurance that a complete freeing of the domestic oil market will lead to the abrogation of the U.S. energy crisis because of decreased consumption and/or more drilling for oil; and it is also claimed that in these circumstances energy users could be expected to diversify their inputs, although of late there seems to be a dearth of suggestions as to precisely how this could be done. The short-run supply of natural gas is at least as uncertain as that of oil, while a meaningful augmentation in the amount of coal available to the industrial countries would require huge investments in such things as transportation facilities, pollution-suppression equipment, and new mines. There is also the matter of recruiting thousands of new workers to a profession that ranks close to the bottom on the popularity scale.

The simple truth is that in the present climate of incomplete information and market imperfections of the type represented by international cartels, we can forget about the supply side of the world petroleum market functioning in the traditional fashion. The background and behavior of OPEC are taken up in chapter 7, but the situation in the United States can be clarified immediately. In 1978, even with the additional oil provided by Alaska, crude-oil output was only 8.7 Mbbl/d, as compared to a 1970 peak of 9.6 Mbbl/d, despite the fact that the 1978 price of oil was between three and five times its 1970 price (depending on whether we are talking about regulated or unregulated oil). More important, with higher oil prices in sight, reserves are falling, which many observers feel is due to decreased exploration: in the lower forty-eight states of the United States, proved reserves have dropped by 40 percent, although production has fallen by only 20 percent. It has even been claimed that for the first time in U.S. history, the drilling boom that followed the 1973-1974 oil-price rises failed to increase domestic oil production. While discussing this topic we should be aware that R.W. Baldwin, the president of Gulf refining and marketing, told a Washington press conference in July 1979 that decontrolling the U.S. oil price would not prevent the continuing decline of production.

At present it would be hard to contradict this opinion; however, it must be admitted that there are several solid arguments in favor of decontrol. Low oil prices have discouraged consumers from saving oil and in general have contributed to a tight world market that invites price increases. Letting the price rise might promote a greater efficiency in energy use; and the increased tax revenue on oil company profits could be used for such things as grants for home insulation or to help finance the infrastructure and pollution-suppressing equipment that would be required if coal were to play a materially larger part in the U.S. energy picture. In my opinion, however, if decontrol comes about, it should take place slowly and under constant monitoring by the authorities, whose primary concern should be the relationship between decontrol and the narrowing of the gap between domestic energy demand and supply. If, as is quite possible, this gap does not show a tendency to crimp after a certain length of time, the experiment should be terminated and written off as a defective tribute to those unenlightened savants who mistakenly believe that the energy crisis was decreed by the governments of the industrial countries.

It might also be appropriate for the reader to recognize that the consumption side of the energy market often seems incapable of reacting in a manner commensurate with the presumptions of a large slice of the academic cognoscenti. In the short run there is only one way for the industrial countries to put their energy house in order—reduce demand. But if we examine the situation as it has developed over the past year, it seems that an overall oil shortage of only 2 percent threw the energy market into a panic. The argument in this book is that this state of affairs was inevitable, given the nature of energy and its relationship to economic progress. If we examine the world production of energy from the middle of the nineteenth century to the present, we see that this is the *only* major industrial factor of production whose output did not decline during the six or seven major recessions in that period, and this includes the three major economic traumas of the present century—the post-World War I recession; the Great Depression, which in many countries involved unemployment rates of 15 to 20 percent; and the post-World War II recession. [Peterson and Maxwell (1979) examine this issue in some detail.] There have been changes in the growth rate of energy use and recently some increase in the efficiency with which this vital input is used in industrial processes; but even now, when the apocalypse might be just over the horizon, almost every industrial country is registering a steady, if slow, increase in its yearly consumption of energy.

Energy and Economic Growth

We begin this section by saying something about the *real price* of oil, which is the *money price* deflated by a suitable index, such as the consumer price index. What the real price indicates is the purchasing power of a unit of oil in terms of consumer or industrial goods. For example, if an individual were to receive a salary in the form of a barrel of oil per month and the real price of oil were to fall by one-half the day before payday, it would mean that the employee could purchase only one-half as many consumer goods with this oil as was possible the previous payday. Conversely, sellers of consumer goods would receive twice as much petroleum for each unit sold of these products. In some places the real price of oil is referred to as the "terms of trade of oil," and a permanent OPEC complaint is that these terms of trade are falling because the decline in the value of the dollar, the monetary unit in which oil is priced at present, "outweighs" the increase in the price of oil.

In the United States, from 1950 to 1973, the real price of *energy* declined at a rate of 1.8 percent a year, while real gross national product (or the *real* value of goods and services produced in the United States) was increasing at 3.7 percent a year. Energy consumption was increasing at 3.5 percent a year, and of this oil was the fastest-growing component. The situation in the United States, moreover, was typical for the entire OECD, where during the 1960-1973 period the demand for energy rose about 1 percent for every 1 percent rise in gross national product (GNP). (Here it should be noted that the International Energy Agency of the OECD predicts that by the year 2000, the energy/GNP elasticity in the industrial world will have decreased to 0.8.) Since 1973 the *average* energy/GNP ratio in the industrial countries of Western Europe has been practically unchanged, while the *marginal* ratio in the United States may have decreased somewhat. On this point it should be appreciated that the energy/GNP ratio in the United States seems to have been declining relative to that of the other industrial countries since the early 1950s for two reasons.

First, although per capita incomes in countries such as the United States, Sweden, Germany, and so forth are now fairly close, per capita energy consumption in the United States is four times that of Germany and twice that of Sweden. It thus seems realistic to assume that much of this energy use in the United States is sheer waste which, as industrial and consumption activities are made more efficient, tends to be decreased. Second, although the *incomes* of many of the OECD countries have increased rapidly, their per capita wealth (or accumulated inventory of buildings, machines, consumer durables, and so on) is still inferior to that of the United States; thus as they tend to acquire a U.S. lifestyle, their energy consumption accelerates.

To get a better insight into this situation, the reader can consider the consumption of air-conditioning services in the United States and Western Europe. Air conditioning has been a quite common phenomen in the United States for a number of years, and with the arrival of the energy crisis restrictions are being placed on its use in public buildings, and increases in the cost of electricity will probably cause many private users to accept a larger amount of discomfort during the summer months. At the same time, air conditioning is just being introduced on a large scale in many European countries, and since as yet it represents only a small percentage of the total demand for energy, even a large decrease in its use would not amount to much in the overall energy picture. Similarly, the relatively poor fuel economy of the U.S. private-transportation sector implies a high potential for direct energy savings without having to substantially decrease either the number of vehicles or the distance they are driven, since apparently drivers are switching to smaller and lighter automobiles. On the other hand, in Europe, where small cars are already the rule, energy savings can come about only through restrictions on speed or distance traveled, and because of political considerations many governments are hesitant to resort to these measures. Figure 2-1 shows the money price of oil in U.S. dollars per barrel. Note in particular that this price is the price of Arabian Light oil, which serves as a *marker* crude: Other oils, having different properties, are priced by using this oil as a reference. Libyan oil, for instance, with a different

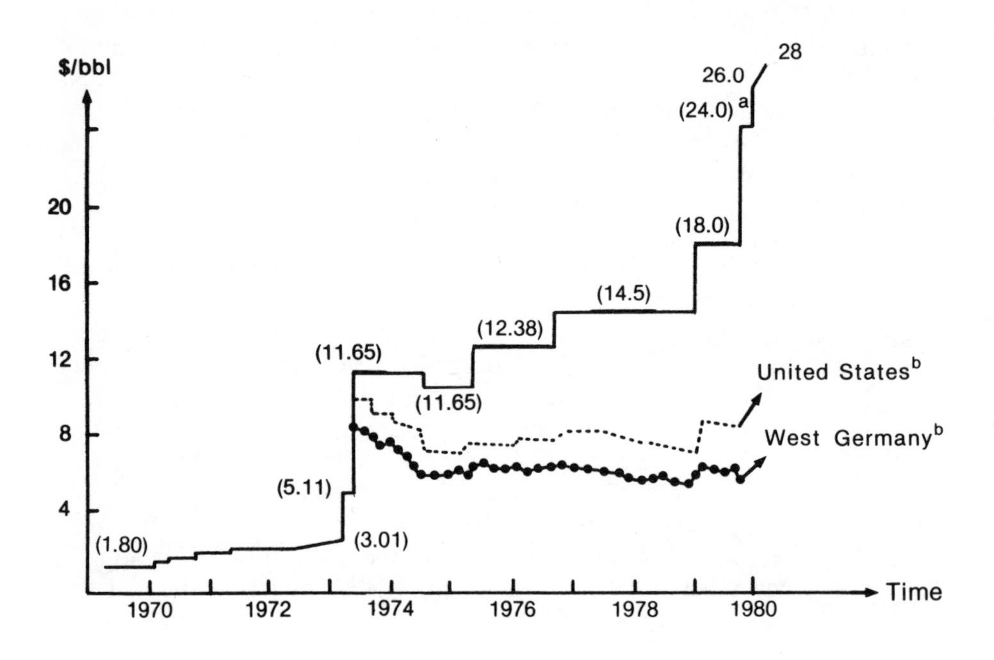

[a]Figures in parenthesis are the posted price of Saudi Arabian crude oil.
[b]The "real price" of oil to the United States and West Germany, calculated with 1972 as the base year.

Figure 2-1. The Money Price and the Real Price of Crude Oil

sulfur content, would always tend to cost more than Arabian Light. Also shown in the figure is the real price of oil to the United States and West Germany, calculated with 1972 as the base year.

Several things in this figure deserve a comment. The first concerns the way in which the real price of oil has shown a tendency to decline almost immediately after a rise in the money price. It could be claimed, in fact, only this phenomenon permits the industrial countries to purchase the oil they require to maintain aggregate rates of economic growth in the vicinity of traditional levels. As a result, an attempt by oil exporters to stabilize the real price of this commodity (through, for example, indexing the money price of oil to the general inflation rate in the oil-importing countries) could have ruinous consequences for the economies of the major oil importers, in terms of either unemployment or inflation or both.

Then, too, the reader should note that the real price of oil has fallen to a greater extent in Germany than in the United States. This can be explained by the appreciation of the German mark relative to the dollar, which in turn was largely caused by the United States drastically increasing the world supply of dollars, in order to pay for its imports of oil, and the recipients of these dollars exchanging them for other currencies, in particular the mark. Thus while German goods become less competitive on world markets because of the rise in the value of the mark, Germans are able to compensate because they can purchase oil and other industrial inputs at a lower price than many of their competitors and thus hold down the price at which they offer their exports. The same phenomenon, in reverse, helps explain why the depreciation of the dollar has not made U.S. goods more attractive to foreign buyers. (Here I can mention that the value of German exports recently surpassed those of the United States for the first time.

Next, in keeping with the exploratory nature of this chapter, we can try to get a deeper perception of the indispensability of energy. As explained in Banks (1977), economic progress often can be boiled down to the introduction of increasingly energy-intensive equipment that eliminates less productive jobs or factors of production, or both. This raises the wage or rental of other inputs, to include various categories of labor, which increases the demand for new types of goods and services, resulting in investment that leads to the reemploying, at higher pay, of individuals freed from less productive work. In many industrial countries this process has also resulted in a high rate of increase in social progress, since it not only satisfied the wage and profit demands of the so-called productive sectors of the economy, but also led to an increasing tax intake that could be used for such things as pensions and welfare payments.

In the period following the 1973-1975 oil-price increases, there was a striking downturn in physical investment in the industrial world. This can be seen for several OECD countries in figure 2-2, which also shows what happened to real output in the OECD countries. The cause of this unhappy spectacle lies in the fact that the cost of energy (like that of wages and salaries) is an operating or variable cost and, for a given value of output, if this increases, then the revenue accruing to capital (machines and structures) or to the owners of capital, which is the same thing, is decreased. In this situation we speak of a decreased *return* or *yield* to capital, and thereby the incentive to purchase more capital is reduced. This fall in yields, or "own rates of return" (which in some textbook cases can be taken as identical to the profit rate) has been well documented by Jorgenson (1978) for the United States. It also seems likely that since the operating cost of many consumer durables was increased by the various boosts in the price of oil, there has been an important decline in the demand for some of these items, and thus profit rates have been further decreased via the medium of falling sales revenues. The plight of certain automobile manufacturers obviously fits into this kind of scenario.

This argument can be made in slightly different terms and expanded

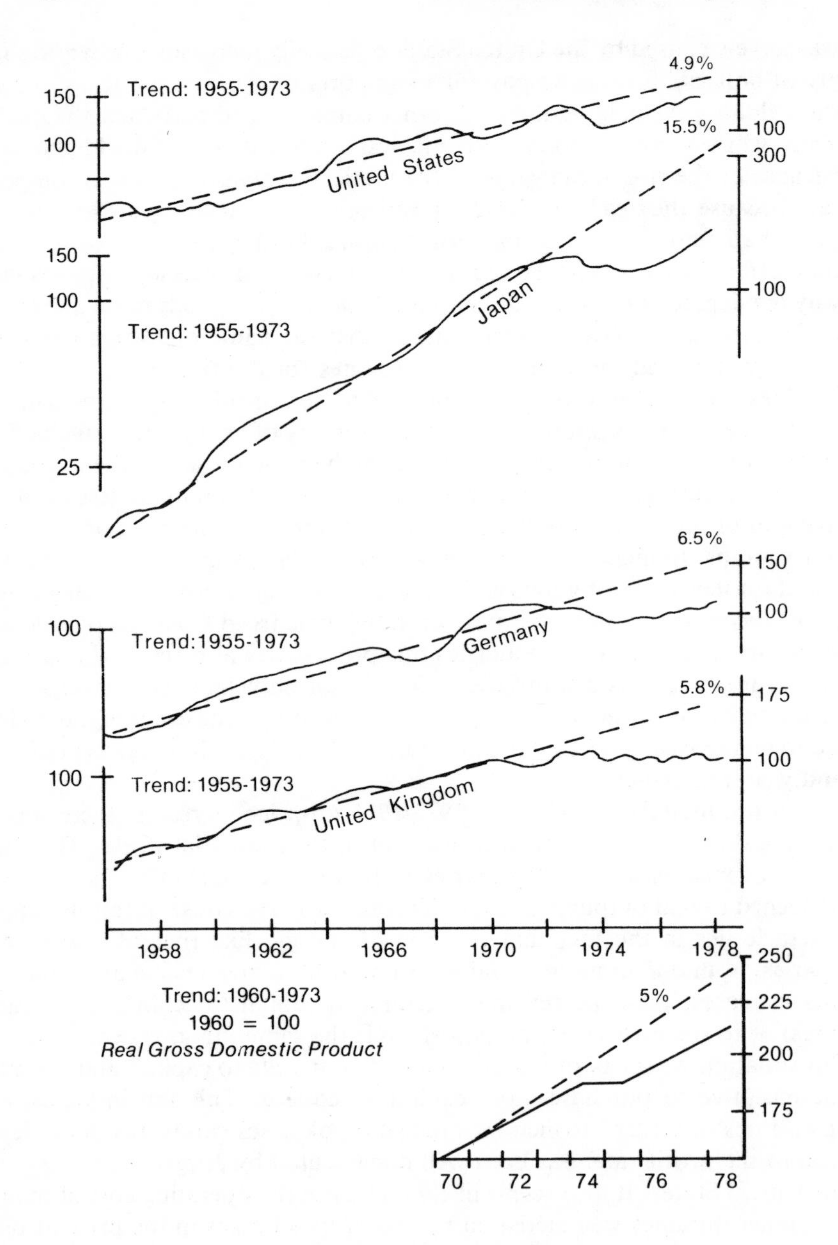

Figure 2-2. The Real Growth of Nonfixed Residential Investment in Four Countries (1970 = 100), including Public-Sector Investment; and the Development of the Total Real Gross Domestic Product for OECD Countries

somewhat. For the same reasons that the oil-price rises have inhibited investment, they have also tended to make a portion of the existing capital stock obsolete: a great deal of equipment is simply too expensive to operate at the new high energy prices. Furthermore, according to Artus (1979), the 1973-1974 oil-price rises caused a reduction in the level of potential output in all these countries between 1974 and 1977 by about 0.6 percentage point per annum. The mechanics of this operation are clarified in chapter 6, where the impact effect of the oil-price rise is shown by using the supply-demand apparatus of elementary economic theory; also see the chapter appendix. But on the basis of the previous exposition, most readers should already have a good insight into the effect of oil-price rises on production costs and the willingness of producers to add to the stock of physical capital. They might also be aware that increases in the price of oil help raise the general price level. This in turn causes a reduction in *real incomes* (which can be defined as *money income* divided by a consumer price index), which tends to reduce the overall demand for goods and services. We can also expect a curtailment in the demand for many products produced in the industrial countries because OPEC oil revenues are not spent immediately after they are earned or, if they are spent, there is a shift in the pattern of demand. To whet the reader's interest at this point, the OECD believes that every 10 percent increase in oil prices can lead to a decrease of 0.3 percent in world industrial output, while increasing world inflation by as much as 1 percent. This last increase includes a direct, or cost-push, component, as well as an indirect component caused by a rise in other energy prices and price rises induced by wage increases resulting from cost-of-living agreements and other employee actions to resist the fall in their real incomes. (See also Noel Uri's (1980) econometric analysis of this topic).

Finally, the fall in investment can be expected to lead to a definitive fall in aggregate rates of economic growth (if this is not already the case) and to move the rate of growth of productivity onto a lower trend. This is so because with a diminished rate of growth of real output—and a larger part of production being used to pay for the growing oil bill—the labor force as a whole is going to have less equipment to work with unless (and this is unlikely) greater savings can be generated out of a smaller real disposable income. The result of these developments will be a definitive fall in the real wage of many employees and/or rise in unemployment, and as is evident from figure 2-2 and table 2-1, the foundation for these outcomes has already been laid.

Before we look at the direct effect of the energy-price increases on consumers, a further aspect of the above exposition needs to be broached. This concerns the possibility of repelling some of the more unsavory effects of the 'energy crisis' by substituting workers for energy-intensive equipment. A former Minister of Energy in Sweden has expressed his deeply felt belief

Table 2-1
Changes in Productivity, Investment, and Gross National Product
(percent per year)

Country	P^a	I^f	B^k	P^b	I^g	B^l	P^c	I^h	B^m	P^d	I^i	B^n	P^e	I^j	B^o
United States	1.8	4.5	4.1	0.1	−6.4	2.0	1.3	−12.5	3.8	−0.3	6.2	2.7	0.8	6.6	3.0
Canada	2.4	5.7	5.6	0.6	5.4	3.3	0.7	3.8	3.5	0.5	2.2	3.8	1.5	1.0	4.0
Japan	8.9	14.0	10.3	3.4	−9.4	3.2	3.9	−2.1	5.8	4.3	3.1	5.2	3.8	11.2	6.0
West Germany	4.7	4.4	4.5	3.2	−9.9	1.4	2.9	−4.2	3.0	3.0	5.0	3.5	3.5	6.3	3.5
Britain	3.2	4.5	3.0	0.8	−2.7	0.3	0.3	−1.9	3.0	2.0	−1.2	3.8	1.5	2.5	2.5
France	4.5	7.6	5.4	3.0	0.9	2.6	3.0	−3.2	2.8	3.3	3.7	3.2	3.3	0.1	3.5
Italy	5.4	4.1	4.5	1.1	3.5	2.0	0.6	−13.0	2.2	2.0	1.9	3.8	3.0	−1.3	3.5

Source: Annual Report of the Bank for International Settlements, 1978, 1979; and OECD Economic Outlook, 1973 to 1979

P = Productivity	*I = Investment*	*B = Gross national product*
a: 1964-1973	f: 1960-1973	k: 1960-1973
b: 1974-1978	g: 1974	l: 1973-1977
c: 1977	h: 1975	m: 1978
d: 1978	i: 1976	n: 1979
e: 1979	j: 1978	o: 1978-1983

that the energy-price rises will increase employment; utilizing his rich fantasy, he sketched for his constituents a low-energy, high-welfare world of expanding opportunity and social harmony. However, despite these fanciful prognostications, present evidence indicates that a low-energy community will be a low-welfare community.

To examine this issue, let us consider a situation in which some work is being performed with energy-intensive equipment, such as electric drills, that could also be performed with picks. To make the example more specific, consider the breaking of rock in preparation for laying the foundations and constructing the cellar of a house. If an upswing in energy prices leads to a substitution of picks for electric drills in order to remove the same quantity of rock in the same time, then employment does indeed rise, but real wages should fall—and in the long run they will fall. The reason is that the ability of the individual rock remover to pay for his or her daily bread and holidays on Majorca is determined on the basis of individual productivity. This in turn is measured by the amount of rock that can be pulverized and carted away per time period, and, as postulated above, the same amount of rock is now being divided between a larger number of people. Moreover, if rock crushers graduated from electric drills to picks but the same number of people were involved, wages would still fall. This is because without the energy-driven drills it would take longer, much longer, to remove the same quantity of rock, and if the rate of interest is positive, time can be translated directly into money. The upshot of this drama is that since wages in most industrial countries are generally immobile downward, the rock breakers would not graduate to picks but simply end up on the dole, their drills would descend to the category of obsolete capital referred to earlier, and, at least for the time being, the potential homeowner would continue to reside in a tent.

The basic issue considered in the last paragraph is the substitutability—or cross substitutabilities—of capital, energy, and labor. If two items are substitutes, then a fall in the demand for one will lead to an increase in the demand for the other; Pepsi Cola and Coca-Cola can be thought of as substitutes. However, if two items are complements, then a fall in the demand for one leads to a fall in the demand for the other; automobiles and gasoline are an example. The contention is that energy and capital are complements, and together they are substitutes for labor. As a *short-run* phenomenon, this complementarity is almost universally accepted: the rise in the price of energy causes a decrease in the use of capital (that is, electric drills) and, if the same production (in the same time) is to be carried out, a rise in employment (because much labor equipped with picks substitutes for a few workers equipped with energy-using drills). But what about the long run?

Some economists claim that in the long run, capital and labor are not complements but substitutes. If this were so in the above example, it would indicate that eventually someone could manufacture a drill having all or most of the features of the original drill, but which would have only a fraction of its energy requirements. Technically this might be possible if the drill were lighter (for example, if it were made of plastic or some very light metal) and if the bearings and other antifriction elements were more efficient. Bringing forth these materials and components would cost a great deal of money, but in the long run it might certainly happen if we did not run into thermodynamic and similar limits. Developmental and production expenses would then be reflected in the price of our new drill: it would cost more than the previous drill, but, by the same token, its operating expense would be lower since it would not require as much high-priced energy. (Another possibility involves investing capital in finding and producing a completely new source of energy. In these circumstances capital does not substitute directly for energy, but since it could lead to a significant increase in the availability of energy, it has almost the same effect.)

Believing that this type of tradeoff can continue indefinitely is consistent with believing that capital is a substitute for energy in the long or very long run. As far as I am concerned, no available evidence offers conclusive proof that this is true *or* not true. But even if it is true, I regard it as completely irrelevant because the "short run" could mean a very long period while the "long run" probably involves a *minimum* of twenty years. In my books *Scarcity, Energy, and Economic Progress* and *The International Economy: A Modern Approach*, in which I make a point of emphatically insisting that capital and energy are complements, I am referring to the short run. Where energy is concerned, I am inclined to take the Keynesian position that economic problems are first and foremost a short-run concern, and the long run is a much less interesting proposition, particularly when we realize that it may have to be measured in decades.

Before leaving this topic, I want to comment on the econometric studies which purport to show long-run capital-labor substitutability *or* complementarity. As bad luck would have it, these investigations have placed some of the leading econometricians and economic theorists in North America at opposite poles, which naturally poses a quandary for those of us who have served only as menials in the econometric wars. But rather than avoid a commitment, let me say that although I believe in the march of science, I do not believe in it enough to expect that our salvation in the present situation will come from redesigning existing structures and machinery to use substantially less energy: the only way out is to find new sources of energy or to change life-styles. *If* energy prices were to increase by even a large increment and then remain stable, I certainly think that new equipment would eventually be developed with lower energy requirements; but with the game being played according to the present rules, the energy-price rabbit can always run faster than the dogs. I also feel that the work of Berndt and Wood (1977) suggests a fundamental dilemma for any researcher trying to derive unambiguous results about capital-energy substitutability or complementarity. These characteristics (substitutability and complementarity) are influenced by the presence or absence of other factors of production (such as nonfuel minerals), which has always been clear. But in constructing a model in which the object is to derive numerical results, one must ask how many of these factors should be included in the investigation, as well as how they should be included, since the presence or absence of a single factor could drastically alter conclusions. There is also the matter of which data sets should be used and the possibility of obtaining meaningful data for some of the excluded factors of production if it is felt that they should be included. Personally, I do not expect to see any of these puzzles solved until long after the energy crisis has subsided and been replaced by more acute torments.

Finally, in the interests of rounding out the discussion, we can note some of the direct effects of energy price rises on the individual consumer. To begin, let us recognize that the immediate effect of an energy price rise is a fall in the utility of many durable goods when taken in relation to their cost. Many people have purchased automobiles, boats, houses, and so on under the assumption that the energy price would either be static or rise only gradually; thus the escalation in the cost of oil has meant an increase in the operating costs of these durables and, all else remaining the same, a higher price for their services. Looking at this matter somewhat more abstractly, we have a situation in which this stock of durables is *inefficient* in the sense that all or a portion of its potential output will not be utilized. The reader searching for some deeper meaning in this arrangement might therefore conclude that given the possibility of shocks of the variety represented by the oil price rises of 1973-1974 and 1979, the price system cannot be expected to allocate resources in the manner prescribed by either our better elementary textbooks or Milton Friedman in his journalistic endeavors. (On this point see the appendix to chapter 8).

Oil Consumption and the United States

One of the principal topics of chapter 6 is the relation between the oil imports of the United States and the value of the U.S. dollar. But rather than hold the reader in suspense, the contention is made now that a major part of the present deterioration in the value of the dollar can be traced directly to the proliferating imports of oil into the United States. Table 2-2 gives some idea of this situation.

A few words are in order about the foreign trade of the United States. In the first twenty-five years following World War II, the United States was more than a match for foreign competitors where the production and selling of both goods and services were concerned. The U.S. merchandise balance is shown in table 2-3, where sales (or, exports) exceeded purchases until 1971. This brilliant record caused some anxiety to the rest of the world because in the postwar reconstruction period the "ideal" situation was

Table 2-2
United States Oil Consumption, Production, and Imports

Year	*Oil Consumption (Mbbl/d)*	*Oil Production (Mbbl/d)*	*Net Imports Mbbl/d)*	*Value of Imports (Billions of Dollars)*	*Value of Imports as a Percentage of Total Merchandise Imports*
1970	14.70	11.30	3.16	2.80	7.00
1973	17.31	10.94	6.03	8.00	11.40
1974	16.65	10.47	5.89	25.60	24.70
1975	16.32	10.01	5.85	26.20	26.70
1977	18.43	9.87	8.57	43.80	28.90
1978	18.82	10.27	7.87	40.40	23.00
1979	18.45	10.10	7.85	59.00	27.90
1980[a]	18.45	10.25	7.70	80.00	35.00

Source: U.S. Department of Energy; *Congressional Research Service Project Interdependence: U.S. and World Energy Outlook Though 1990.*
[a]Estimated.

ostensibly the United States importing more than it exported and thus generating a supply of dollars for foreign countries which would have permitted them to buy (mostly from the United States) the equipment they needed to rebuild their economies. In fact, until the mid-1960s the "dollar shortage" was a favorite subject for many economists. For a full account of this topic and its extensions, see Banks (1979a) and the references cited there.

In addition to the surpluses on the merchandise account, there was also a large demand for dollars in order to buy U.S. stocks and bonds, to purchase property in the United States, or simply to hold in bank accounts in the United States. To a certain extent this was countered by U.S. direct investment abroad as U.S. firms established affiliates and bought their way into foreign businesses; but on the whole the U.S. capital account, like the merchandise account, generally showed a surplus. A great deal of this was explained by the fact that New York was—and perhaps still is—the financial center of the world; and even when London regained some of its prewar luster, there was still a huge demand for the services provided by various New York banks and financial institutions.

Where, then, were the negative entries that made the accounts balance? The answer lies in official U.S. lending, gifts, grants, and aid. The Marshall Plan, which provided a large part of the money needed to get Europe started on its present round of prosperity, was a major source of deficits; and later, as the Cold War threatened to become warmer, there were extensive, and to a certain extent unnecessary, charities to some of the more incompetant and corrupt governments in the less developed world.

The turning point in the history of U.S. foreign economic relations came in the middle of the 1960s, when the economic power of Western

Table 2-3
U.S. Trade Balances and Value of the Dollar and Some Macroeconomic Data for Selected Countries and the OECD
(Trade balances in billions of U.S. dollars, and inflation and growth rate in percent)

Year	U.S. Merchandise Balance	U.S. Current-Account Balance	Value of the Dollar[a]	Annual Inflation Rate				Growth Rate OECD
				United States	West Germany	Japan	OECD	
1950	+1.1	−2.1	—	—	—	—	—	—
1951	+3.1	+0.3	—	—	—	—	—	—
1952	+2.6	−0.2	—	—	—	—	—	—
1953	+1.4	−1.9	—	—	—	—	—	—
1954	+2.6	−0.3	—	—	—	—	—	—
1955	+2.9	−0.3	—	—	—	—	—	—
1956	+4.8	+1.7	—	—	—	—	—	—
1957	+6.3	+3.6	—	—	—	—	—	—
1958	+3.5	−0.1	—	—	—	—	—	—
1959	+1.1	−2.1	—	—	—	—	—	—
1960	+4.9	+1.8	—	—	—	—	—	—
1961	+5.6	+3.1	—	1.1	2.3	5.3	1.8	—
1962	+4.5	+5.2	—	1.2	3.0	6.8	2.4	5.3
1963	+5.2	+3.2	—	1.2	3.0	8.5	2.6	5.5
1964	+6.8	+5.8	—	1.3	2.3	3.9	2.4	6.0
1965	+5.0	+4.3	—	1.7	3.4	6.6	3.0	5.3
1966	+3.8	+1.6	—	2.9	3.5	5.1	3.4	5.5
1967	+3.8	+1.3	—	2.8	1.4	4.0	3.1	3.8
1968	+0.6	−1.3	—	4.2	2.9	5.3	4.0	5.5
1969	+0.6	−2.0	—	5.4	1.9	5.2	4.8	4.7
1970	+2.6	−0.4	—	5.9	3.4	7.7	5.6	3.1
1971	−2.3	−4.0	1.086	4.3	5.3	6.1	5.3	3.8
1972	−6.4	−9.9	1.086	3.3	5.5	4.5	4.8	5.5
1973	+0.9	+0.4	1.206	6.2	6.9	11.7	7.9	6.3
1974	−5.4	−5.0	1.224	10.9	7.0	24.3	13.1	0.5
1975	+9.0	−11.6	1.171	9.2	5.9	11.9	10.8	−0.4
1976	−9.3	−1.4	1.162	5.8	4.5	8.3	7.8	5.2
1977	−31.1	−15.3	1.215	6.5	3.9	8.1	7.8	3.7
1978	−32.0	−15.2	1.303	9.0	2.6	3.8	6.8	3.9
1979	—	—	—	13.0	—	—	—	3.3
1980	—	—	—	—	—	—	—	1.0[b]

Source: IMF and OECD statistics.
[a]In dollars per special drawing right. Sixteen major currencies are averaged to compute the value of Special Drawing Right.
[b]Estimated.

Europe was fully restored and Japan was well on its way to establishing the most productive economy in the world. President Kennedy's small initial military commitment to Vietnam became the entrée to a full-scale war that completely altered both the political and the economic situation of the United States on this troubled planet—perhaps forever. Among other things, it was found necessary to finance a sizable part of the official U.S. expenditures during this period with the printing press. As a result, large quantities of dollars eventually overflowed into Western Europe and Japan where harried businessmen, bankers, and governments found, to their growing disgust, that instead of having too little of this once precious lucre, they now had too much. Moreover, some problem was experienced in getting them back to the United States because the attractiveness of U.S. goods was no longer such that foreigners automatically set sail for North America when they wanted to shop for machinery, vehicles, or electronic gear.

It should also be stressed that the war in Vietnam distorted economic development in the United States by moving resources into the production of military goods that should have been used to modernize the U.S. industrial capital stock, rejuvenate the educational system, and make the energy investments which would have obviated some of the discomforts associated with the deteriorating U.S. energy position. (At this point the reader should be aware that a large part of German, and particularly Japanese, economic success can be explained by their comparatively low level of military spending.) The war also raised moral issues that led to serious social conflict. Such socially disrupting practices as drug use became more widespread, especially in so-called respectable circles, and the celebrated Protestant work ethic (which was actually the U.S. work ethic) was severely diluted. Although it is not widely realized, this issue lies close to the heart of the unprecedented decline in productivity that now characterizes an uncomfortably large part of economic life in the United States.

When the war ended, the U.S. dollar was in distinct oversupply, but some exchange-rate adjustments were underway which, with a bit of luck, might have succeeded in bringing the foreign accounts back into balance. (At least, the merchandise surplus of 1973 seems to suggest this possibility.) But then the October war among Egypt, Syria, and Israel began in the Middle East. In the last part of 1973 and the first part of 1974, the price of oil increased by a factor of four. Since that time it has continued to move up, albeit slowly, until 1979 when, in a rapid series of ad hoc moves (later ratified by the OPEC directorate) the price of OPEC oil increased by a total of almost 100 percent over the December 1978 price.

The present dilemma of the dollar originated with a loose-money policy designed to ensure that individuals and firms in the United States that require dollars to pay for imported oil can get them. Many of these dollars then move, by way of the Euromarket, from the oil-producing countries to the rest of the world where, because of their number and the fact that the world is now on a "dollar standard," they pose a permanent threat to international financial stability. The sad thing about all this is that with unwanted dollars piling up in their central banks at an unprecedented rate, the political leaders of Western Europe and Japan have not been in the mood to join their counterparts in North America and Australia in a serious attempt to design a viable energy strategy, even though it seems clear that with the ample energy materials and scientific knowledge present in those countries, energy independence could be a lot closer than many people realize.

Since I have already reviewed some of the better-known miseries associated with the oil price rises, I close this section by saying a few words about one that is less well known. This has to do with the exporting of industrial goods in order to pay for oil at just the time when this equipment is badly needed in its countries of origin. Specifically, the problem is one of having to give up a growing share of a decelerating industrial production in order to buy a product (oil) whose productive power, as an industrial input, is virtually constant. Apparently the decisionmakers in the industrial countries have chosen to overlook this rather unpleasant aspect of the energy crisis, either because if they did notice it, there is nothing they can do about it or because they have mistakenly become convinced (or convinced themselves) that paying for oil with machinery costing ten or twenty times as much in real resources is actually good business. Table 2-4 provides some indication of the development of OPEC trade.

One final word for the reader who has not yet grasped the full significance of this issue. A year ago the U.S. Corps of Engineers was reported by *Fortune* magazine to be doing more work for the government of Saudi Arabia than for the U.S. Armed Forces in Germany. Given that the price of oil has now doubled, perhaps in the near future they will not be able to afford to work anywhere except in the oil-producing countries. Put less vividly, OECD countries are now paying 3.8 percent of their collective GNP for oil imports, as compared to 1.2 percent in August 1973.

Oil and the Future

In case the reader has inferred from the first part of this chapter that one of the purposes of this book is to prophecy the decline and fall of our civilization, let me emphasize that where energy is concerned, optimism is almost certainly a better position than pessimism—at least as a long-term proposition. Although progress does not seem to be rapid at present, we are in transit from an arrangement featuring low cost but limited supplies of fairly inexpensive energy resources (such as oil), to energy supplies requiring fairly high capital (or, equipment) costs. But once the investments in the new energy technology are complete and such things as nuclear and synthetic fuel installations are in operation on a large scale, the average cost of energy could be as low as, or even lower than, those being experienced now.

Still, some care must be taken in the choice of the new technology. Stobaugh and Yergin (1979), working and publishing under the name of the Harvard Energy Project, have expressed the noble idea that solar technology and conservation should serve as the underpinning of an enlightened energy policy in the United States and, presumably, any country facing an energy deficiency. They also give some indication of how this is to come about, although it appears that neither their conclusions nor their

Table 2-4
Foreign Trade of the OPEC Countries, OPEC Trade by Categories of Goods, and OPEC Trade with Groups of Other Countries

Foreign Trade (Millions of U.S. Dollars)	*1970*		*1973*		*1974*		*1977*		*1978*	
	X	*M*	*X*	*M*	*X*	*M*	*X*	*M*	*X*	*M*
Algeria	1,009	1,257	1,896	2,408	4,602	4,048	5,810	7,125	5,855	7,868
Iraq	1,095	510	1,955	894	6,972	2,370	9,664	4,052	11,099	3,500
Libya	2,843	554	3,454	1,813	7,127	2,763	9,761	3,782	9,503	5,886
Kuwait	1,691	624	3,321	1,052	9,856	1,552	9,801	4,840	10,543	4,613
Qatar	258	64	619	195	2,015	271	1,986	1,255	2,317	1,000
United Arab Emirates	550	267	1,807	821	6,413	1,705	9,447	4,503	9,049	4,892
Saudi Arabia	2,173	711	7,747	1,975	30,989	2,895	41,210	14,656	37,935	2,852
Iran	2,623	1,662	6,205	3,393	21,554	5,427	24,246	13,750	22,549	19,186
Gabon	144	80	382	160	766	322	1,095	831	1,000[a]	600[a]
Nigeria	1,240	1,059	3,526	1,865	9,216	2,763	11,780	11,095	10,509	11,913
Equador	221	274	358	397	1,135	678	967	1,508	629[a]	1,583[a]
Venezuela	2,598	1,811	4,891	2,810	10,883	4,190	9,548	10,353	9,126	10,373
Indonesia	1,152	1,002	3,211	3,393	7,449	3,842	10,853	13,750	11,643[a]	19,186[a]
OPEC Total	17,588	9,876	39,372	21,177	118,927	32,800	146,198	91,470	141,667	113,452
OPEC Trade by Category as Percentage of OPEC Trade										
Food and Beverage	5.1	10.7	—	—	1.3	13.9	2.0	10.4	—	—
Raw Materials	6.7	3.8	—	—	2.1	4.3	1.9	2.6	—	—
Mineral Fuels	85.9	1.7	—	—	95.3	1.9	94.7	2.2	—	—
Chemicals	0.4	8.4	—	—	0.4	8.1	0.2	0.1	—	—
Machines and Transport	0.3	36.9	—	—	0.2	36.8	0.0	47.0	—	—
Other Goods[b]	1.6	38.5	—	—	0.7	35.0	1.4	37.8	—	—
Total	100.0	100.1	—	—	100.0	100.0	100.0	100.0	—	—

OPEC Trade with Groups of Other Countries as Share of World Trade

Industrial Countries	4.4	2.5	—	—	11.3	3.5	9.9	6.2	8.5	6.2
OPEC	—	—	—	—	—	—	0.1	0.1	0.1	0.1
Other Developing Countries	1.0	0.3	—	—	2.7	0.6	2.8	0.8	2.2	0.8
Centrally Planned Countries	—	0.3	—	—	0.2	0.3	0.2	0.4	0.2	0.5
Total	5.7	3.1	—	—	14.2	4.4	13.0	7.6	11.0	7.6

Source: IMF Financial Statistics; UNO Statistical Yearbook (Various Years).

Note: OPEC trade by categories of goods is in percentage of OPEC trade. On the other hand, the foreign trade of OPEC by groups of countries is in terms of the share of world trade. For example, in looking at the bottom part of the table, we see that OPEC exports (by value) have gone from 5.7 to 11 percent of world exports in 1970-1978, and imports from 3.1 to 7.6 percent.

[a]Estimated.

[b]Mostly other manufactures.

specifications for the implementation of these conclusions is being taken seriously at present. In fact, given that about six years of well-paid contemplation went into this scholarly venture, it appears that from a cost/benefit point of view this project showered alms on academics in a manner that would be considered extreme even by the wasteful standards of Swedish economic research. Although it is true that the Stobaugh and Yergin study contains some important observations and certainly deserves to be examined, it so happens that what is taking place in the real world does not make a very strong case for the "soft energy" options it advocates.

It seems to be a fairly well-established empirical fact that about thirty years are needed for a new energy source to capture 10 percent of the energy market—assuming that the new technology is available in commercial form. Thus if we were to make the (unrealistic) assumption that a wide range of solar devices will be commercially feasible from about 1980, the year 2010 would show a consumption of solar energy at a level between one-fourth and one-half percent of that envisioned by Stobaugh and Yergin. Of course, a crash program initiated by governments induced to shift their attention from familiar trivia to energy matters *could* achieve a more satisfactory result, but as far as I know, there are no such programs on the horizon.

What about conservation? The director general of the International Energy Agency has stated that if the energy/GNP ratio could be reduced by one-eighth, then 10 Mbbl/d of oil could be saved by the year 2000. This ratio is falling, and it is not inconceivable that by the year 2000 it will have fallen by one-eighth in the industrial countries. The present shift in world income and the increase in global population would ensure, however, that a part of this gain is erased. Similarly, it has been calculated that if annual economic growth were reduced for 3.4 to 2.4 percent for the industrial world, 20 Mbbl/d could be saved by the year 2000. Roughly speaking, though, this would mean a minimum unemployment rate in most industrial countries of 8 to 10 percent in the same year, in addition to an annual loss in world production of $1 trillion to $2 trillion. A general decrease in the level of real consumption per head could also be expected throughout the industrial world. However, the political ramifications of this kind of scenario were not taken up by the secretary general although had he bothered, I feel sure that he would have found them disturbing.

The basic impasse where conservation is concerned is as follows. The U.S. way of life is such that a movement away from the automobile and back to public transport and more concentrated cities is not conceivable for psychological, sociological, and other reasons that should be more than obvious to anyone reading this book. It is equally true that such things as cheap oil and personal transport are regarded as a basic democratic right in all European and Europeanized countries. Here the reader would do well to remember that John Maynard Keynes, whose popularity as an economic seer and innovator appears to be rising to new heights in tune with the breakdown in the explanatory power of some of his theories was able to agree that in time of war every citizen of a country is obligated to subordinate himself or herself to the duly constituted authorities, although *he* chose not to register for compulsory military service in 1916 at a time when the British Army was being systematically decimated in Flanders. Keynes gave as his reason personal conviction, when actually what he meant was personal convenience. The same thing applies now. In the industrial countries today it is almost impossible to find an intelligent person who does not feel that more effective conservation measures should be introduced immediately; but very often this means measures introduced for *other* people.

Now we can examine how OPEC views this coming and going on the conservation front. For various reasons, many OPEC countries are now making it abundantly clear that they simply are not going to produce as much oil as many politicians in the industrial countries feel they should produce. What they are saying is that the industrial world is going to have to conserve oil whether it wants to or not. If industrial countries choose to do otherwise, they are going to have to pay dearly for their extravagance (although, as is argued later, they are going to have to pay dearly anyway). If we take a country such as Saudi Arabia, less than a year ago so-called informed opinion was claiming that by 1990 Saudi Arabia would be producing between 18 and 20 Mbbl/d of oil. At present a ceiling of 12 to 13 Mbbl/d is the absolute best that could be hoped for, and some experts have started to claim that the maximum sustainable output of that country is 10 Mbbl/d. At the same time, there is purportedly an important body of opinion within Saudi Arabia which believes that Saudi Arabian economic and social goals can best be met if oil production is not permitted to exceed 8.2 Mbbl/d.

Since it was a widely advertised belief that a Saudi Arabian output of less that 14 to 16 Mbbl/d by 1985-1990 constitutes very bad news for the industrial world, the reader should try to appreciate that for the Saudis to get their production up to 16 Mbbl/d, they would have to invest a minimum of $15 billion (U.S.) and, it now appears, this production could be maintained for only five to ten years. Thus the major part of the productive life of these new investments would be wasted. In addition, Saudi Arabian economists have started to ask why they should exchange a large part of their valuable oil for financial assets with a dubious future value. If the money value of the oil imported by the United States continues to increase at the present rate and the stock of U.S. dollars in foreign hands continues to expand the way they are expanding just now, then we could get an accelerating diversification out of that currency and a collapse of the financial system based on the dollar. This system embraces the entire noncommunist world, and although they may not realize it, some of the oil-producing countries have as much of a stake in it as Wall Street bankers.

We can conclude this chapter by taking a short look at some political, quasi-political, and "ivory tower" economic measures to alleviate the oil crisis and the general economic turmoil moving it its wake since, as Congressman Morris K. Udall makes clear, energy is going to dominate both the politics and the lives of people of the United States (and other industrial countries) for a long time.

The first thing that must be done is to reaffirm President Carter's widely disliked proposition that the energy crisis is going to present the citizens of the industrial world with what amounts to an equivalent of war. War, as we know, means sacrifice, and so the President undoubtedly did not help his political career by his inept choice of terminology, but the energy impasse remains regardless of what the electorate wishes to call it. People in general, and people in the United States in particular, have a fundamental belief in cheap, abundant energy—particularly when they are behind the wheel of their automobiles. But the days when this belief can be translated into fulfillment may be gone. While we are on this subject, it should be noted that the possibility has been broached of military intervention against OPEC, particularly in the Middle East. United States marines and paratroopers, lavishly supported by low-flying aircraft, could not only seize a large segment of the oil fields in that part of the world but also, even if these installations were badly damaged, have them back in operation in a short time. Even so, this would probably be the most pointless military enterprise since the French and British invasion of Egypt in 1956, and, if anything, the political and psychological consequences would be immeasurably more grave than those resulting from that previous ill-timed and unnecessary adventure.

The French government, on the other hand, seems to advocate a consumer oil cartel that would hold a 'dialogue' with OPEC. French politicians have always been strong on dialogues, and the setting up of a permanent or semipermanent talkathon with the "oil club" on one side and OPEC on the other might conceivably relieve some tensions. But, by the same token, it might create some, particularly if the United States were involved. There is a growing intolerance of OPEC on several levels of the U.S. scene although, in point of truth, the OPEC directorate has been a paragon of moderation throughout most of its existence. With all due respect to the bureaucrats and diplomats who would become involved, I suggest that if they have any real competence in questions relating to energy, then it is a waste of time to display it in the high-rent surroundings of Parisian conference rooms and restaurants. There are plenty of people in their home countries requiring an extensive education in these matters; in particular, it seems that a surprisingly large number of decisionmakers in the industrial world lack even a basic conception of the dangers associated with an extended shortage of energy.

As for the economic remedies, suffice it to say that the academic community is steadily growing richer in nonsolutions, and so I mention only two. Professor Henrik Houthakker has suggested putting a tariff on imported oil and using the proceeds to invest in alternative energy resources. Energy prices would be higher for the individual consumer, but it could be reasoned that they are going to go up in any case, and Houthakker's point is to provide capital for a new energy technology. If we overlook the unhappy fact that an increase in the price of oil causes a noticible increase in the domestic price level, which is already rising at a record pace in most industrial countries, then in the energy field the problem is to settle on *which* investment to finance of the many that appear plausible. As far as I know, the major oil companies are not short of cash, and several of the largest firms are now making substantial financial commitments outside the energy sector.

In line with his previous contributions, Professor Adelman proposes that entitlements to sell oil to the tune of 200 million tickets worth $2.5 billion should be auctioned off each month. These tickets would be bought anonymously, and their purpose is to provide OPEC producers having an urge to sell under the official price with the means of doing so (since this auctioning would function something like the bidding for certain types of contracts, with the contract going to the lowest bidder). Since Adelman, together with Friedman, was for many years a forceful advocate of the position that OPEC could not maintain the price of oil too far above its marginal cost, admitting that OPEC has significant market power which can be reduced by the assiduous application of some kind of economic manipulation represents progress of a sort. But since the decisionmakers in many OPEC countries consider themselves fully prepared to counter this and similar strategies, I believe that Adelman's proposal should be shelved for the time being and more attention should be paid to suggestions of the Department of Energy (DOE). The DOE submits that the surest way back to economic stability for the industrial world is to consume less oil and install an alternative energy technology as rapidly as possible.

To complement this discussion, see the figures given in table 2-5 for the world production and consumption of oil, as well as the *marginal* energy requirements relative to gross national products for some of the most important industrial countries. A comment which might be appropriate here is that in the case of, for example, the OECD, where the marginal energy requirement is apparently below unity, the average is still slightly above unity, but it the marginal value remains below the average, it will pull down the average. (Another expression for the word *marginal* is *incremental,* and this term refers to a change in some variable of interest).

Table 2-5
World Oil Consumption and Production Data and Marginal Energy Requirements

	1968		*1978*		*Estimated Value of Oil Imports (Billions of Current Dollars)*			*Marginal Energy Requirements*[c]		
	Consumption (Mbbl/d)	*Production Mbbl/d)*	*Consumption (Mbbl/d)*	*Production (Mbbl/d)*	*1978*	*1979*	*1980*	*1960-1969*	*1969-1973*	*1975-1978*
Western Hemisphere	16.9	16.9	24.4	16.8	—	—	—	—	—	—
United States	13.1	10.6	18.3	10.3	42	58	75	0.97	1.07	0.76
Canada	1.4	1.2	1.8	1.6	1	2	2	1.04	1.12	0.69
Mexico	—	0.4	—	1.3	—	—	—	—	—	—
Venezuela	—	3.6	—	2.2	—	—	—	—	—	—
Other	2.4	1.1	3.3	1.4	—	—	—	—	—	—
Western Europe	10.2	0.5	14.6	1.8	—	—	—	—	—	—
United Kingdom	1.8	—	1.9	1.1	4	3	1	0.75	0.67	0.55
Germany	2.1	0.2	3.0	0.1	14	22	24	1.06	1.03	0.96
France	1.5	0.1	2.4	—	11	17	18	0.75	1.30	0.96
Italy	1.4	—	2.0	—	8	13	14	1.63	1.77	1.15
Norway	0.1	—	0.2	0.4	—	—	—	—	—	—
Other	3.3	0.2	5.1	0.2	—	—	—	—	—	—
Middle East	0.8	11.2	1.7	21.3	—	—	—	—	—	—
Saudia Arabia	—	2.8	—	8.3	—	—	—	—	—	—
Iran	—	2.8	—	5.2	—	—	—	—	—	—
Iraq	—	1.5	—	2.6	—	—	—	—	—	—
Kuwait	—	2.4	—	1.9	—	—	—	—	—	—
Abu Dhabi	—	0.5	—	1.5	—	—	—	—	—	—
Other	—	1.2	—	1.8	—	—	—	—	—	—
Africa	0.7	4.1	1.2	6.1	—	—	—	—	—	—
Libya	—	2.6	—	2.0	—	—	—	—	—	—
Nigeria	—	0.1	—	1.9	—	—	—	—	—	—
Algeria	—	0.9	—	1.2	—	—	—	—	—	—

Other	—	0.5	—	1.0	—	—	—	—	—	—
Asia and Australia	4.8	0.8	9.1	2.9	—	—	—	—	—	—
Japan	2.8	—	5.4	—	25	38	42	1.09	0.83	0.43
South and Southeast Asia	1.5	0.8	2.9	2.4	—	—	—	—	—	—
Australia	0.5	—	0.8	0.5	—	—	—	—	—	—
Centrally Planned Economies	5.7	6.8	12.2	14.0	—	—	—	—	—	—
USSR	4.5	6.2	8.4	11.7	—	—	—	—	—	—
Eastern Europe	0.9	0.3	2.1	0.4	—	—	—	—	—	—
China	0.3	0.3	1.7	1.9	—	—	—	—	—	—
Total	39.2	40.4	63.1	63.0	152[a]	218[a]	260[a]	0.97[b]	1.07[b]	0.76[b]

Source: Various International Energy Agency Reports.

[a]Including countries not shown (for example, LDCs).

[b]For OECD.

[c]Marginal energy-GNP ratio, in percent.

Appendix 2A: The Elasticity of Substitution and the Effect of Increased Import Prices

Elasticity of Substitution

The *elasticity of substitution* is defined as follows, where x_1 and x_2 are inputs (for example, capital and labor) and r_1 and r_2 are the rental rates of these inputs (the rental rate of capital and the wage of labor):

$$\sigma = \frac{d \log (x_2/x_1)}{d \log (r_1/r_2)} \tag{2A.1}$$

This can obviously be written as

$$\sigma = \frac{d \log (x_2 - x_1)}{d \log (r_1 - r_2)} = \frac{dx_2/x_2 - dx_1/x_1}{dr_1/r_1 - dr_2/r_2} \tag{2A.2}$$

Let us now assume that we have only an increase in r_2, so that r_2 increases relative to r_1. Then we get

$$\sigma = \frac{dx_2/x_2 - dx_1/x_1}{-dr_2/r_2} = \frac{dx_1/x_1 - dx_2/x_2}{dr_2/r_2} \tag{2A.3}$$

Rewriting, we get

$$\frac{dx_1}{x_1} - \frac{dx_2}{x_2} = \frac{d(x_1/x_2)}{x_1/x_2} = \sigma \frac{dr_2}{r_2} \tag{2A.4}$$

In the two-factor case shown in figure 2A-1 we see that we must have $0 \leq \sigma$. Since the elasticity of substitution is defined on the isoquant (for example, isoquants Q in figure 2A-1), in figure 2A-1*a*, where we have a neoclassic situation, a change in relative prices must lead to a decrease in the use of the factor whose relative price has increased; in figure 2A-1*b*, where we have an input-output isoquant, a change in relative prices does not lead to a change in factor proportions. In other words, the elasticity of substitution is equal to zero. Taking another look at the arrangement in figure 2A-1*a*, where we have substitution, we must have

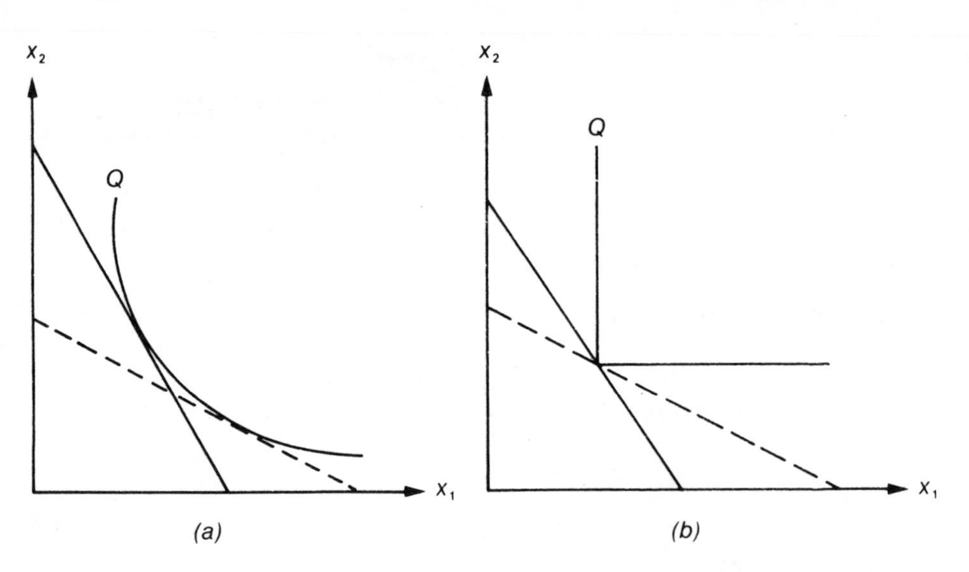

Figure 2A-1. Representative Isoquants and Changes in the Factor/Price Ratio from r_1/r_2 to r_1/r'_2, with $r'_2 > r_2$ and r_1 Constant

$$\frac{dr_2}{r_2} \uparrow \quad \Rightarrow \quad \frac{d(x_1/x_2)}{x_1/x_2} \uparrow$$

Thus $\sigma > 0$. (Although in general σ is defined as negative.)

But when we have three (or more) factors, complementarity among factors is possible. In fact, in the electric-drill example given in this chapter there was complementarity between energy and capital, and together they were a substitute for labor. This means that the sign of the substitution elasticity between capital and energy would be negative. For the best explanation of this matter and an examination (and partial reconciliation) of some opposing points of view, refer to Berndt and Wood (1977).

Effect of Increasing Import Prices

Now we discuss the effect on a model economy of an increase in import prices. The basic intention here is not so much to make a point (which is obvious in any case), but to suggest a simple analytical scheme which other economists might be interested in extending in the direction of less or greater complexity. A simple, two-sector input-output table of the type shown in figure 2A-2*a* describes our economy. In order to keep the exposition simple, it is assumed that there are no interindustry flows, that only one

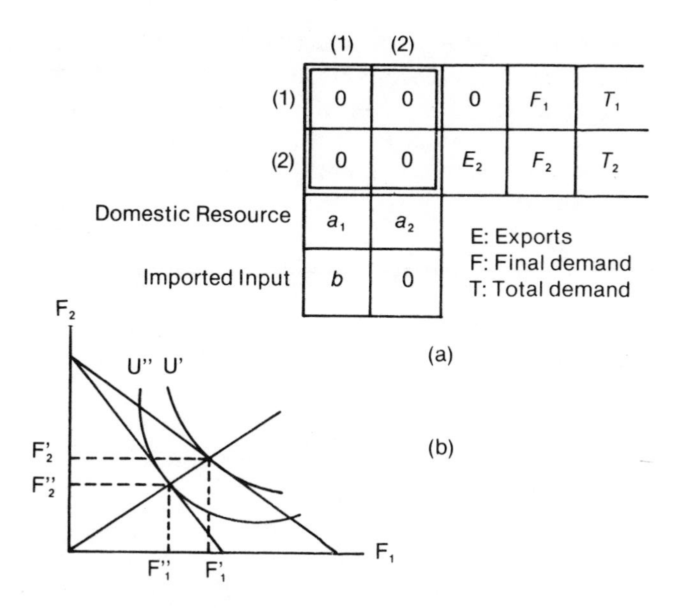

Reprinted by permission of the publisher from Ferdinand E. Banks, *Scarcity, Energy, and Economic Progress* (Lexington, Mass.: Lexington Books, D.C. Heath and Company). Copyright © 1977, D.C. Heath and Company.

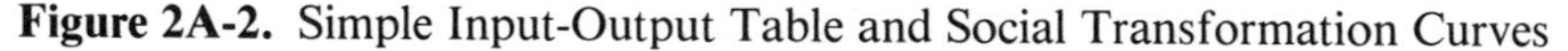

Figure 2A-2. Simple Input-Output Table and Social Transformation Curves

sector requires imported inputs (of oil, for example), and that only the output of sector 2 is exported. There is a final (domestic) demand F_1 and F_2 for both products, and both use some of the domestic resource R as an input in production. Taking P_r as the rental rate of the domestic resource and P_m as the price of the imported input, and T_1 as the total output of sector 1, we get the following input-output relationships from figure 2A-2*a*:

$$a_1 P_r + b P_m = P_1 \quad (2A.5)$$

$$a_2 P_r = P_2 \quad (2A.6)$$

$$b T_1 = M \quad (2A.7)$$

$$a_1 T_1 + a_2 T_2 = \overline{R} \quad (2A.8)$$

Equation 2A.8 gives the total requirement of the domestic resource and equation 2A.6 the price of the export good. It should be noted here that the price P_2 is determined by the price of the domestic resource P_r, which is a kind of internal *numeraire;* P_m is assumed to be fixed abroad; M is determined by final-demand requirements (from equation 2A.7); and, as will be seen, the trade-balance restriction fixes the required exports. In order to put some restrictions on production, there would have to be some restrictions on R—for example, $R \leq R^*$—but that has not been done here. Continuing, we have

$$P_m M = P_2 E_2 \quad \text{(the trade balance)} \tag{2A.9}$$

$$F_2 = T_2 - E_2 \tag{2A.10}$$

$$F_1 = T_1 \tag{2A.11}$$

From equations 2A.8, 2A.10, and 2A.11 we now get

$$\overline{R} = a_1 F_1 + a_2(F_2 + E_2)$$

Similarly we have $E_2 = P_m M/P_2 = bP_m T_1/P_2 = bP_m F_1/P_2$ *from equations 2A.6, 2A.9, and 2A.11. Thus we get*

$$\overline{R} = F_1 \left[\frac{a_1 P_r + bP_m}{P_r}\right] + a_2 F_2 \tag{2A.12}$$

Using equations 2A.5 and 2A.6, we can write equation 2A.12 as

$$F_2 = \frac{\overline{R}}{a_2} - \frac{P_1}{P_2} F_1$$

This can be called a social transformation curve: it shows the tradeoff between F_1 and F_2. We now need some means of choosing a point on this curve. One possibility is to define a utility function $U = U(F_1, F_2)$ and solve the following nonlinear program:

$$U = U(F_1, F_2) = \max$$

$$F_2 = \frac{\overline{R}}{a_2} - \frac{P_1}{P_2} F_1$$

$$F_1, F_2 \geq 0$$

For this simple program the Kuhn-Tucker conditions, with λ the shadow price of R, give

$$\lambda\left(\frac{P_1}{P_2}\right) \leq U_1$$

$$\lambda \leq U_2 \tag{2A.13}$$

If we assume a neoclassical (interior) solution, this becomes

$$\frac{U_1}{U_2} = \frac{P_1}{P_2} \tag{2A.14}$$

Diagramatically this takes the form shown in figure 2A-2*b*. The increase in P_m (the price of oil), *ceteris paribus,* increases P_1 (since $P_1 = a_1P_r + bP_m$). This causes, as the reader can verify, the social transformation curve to swing inward. Thus in figure 2A-2*b* we move from utility curve U' to utility curve U''. This is equivalent to a fall in real income.

The Supply of Oil: An Introduction

This chapter introduces several crucial concepts. The most important is the reserve/production ratio; although first we discuss resources and reserves. Later a few things are said about exploration, but a serious attempt has been made to reduce this phase of our exposition to the bare essentials.

Resources and Reserves

Crude petroleum, or oil (as it is called in this book), is found in small, open spaces (pore spaces) in permeable reservoir rock, and not in large, open caverns, as is commonly thought. An oil field consists of one or more distinct accumulations of this nature which, inappropriately enough, are called pools. In this situation, if sufficient gas is dissolved in the oil and the confining pressure is high enough, a well drilled into oil will release that pressure, and the expanding gas will drive the oil to the surface, sometimes in the form of a gusher. Less dramatically, the buoyancy of oil over water may simply push the oil to the surface. In neither case does oil have to be pumped, at least during the early life of the field. However, the rate of production that can be sustained from any particular reservoir tends to decline as the oil in it is depleted.

If the natural pressure of the field is, or becomes, insufficient, the oil is pumped out. The two arrangements mentioned here (natural release and simple pumping) are called *primary production,* and according to Flower (1978), in the United States in 1977-1978 the average amount recovered by primary production was 25 percent. In fields where primary production results in an inadequate yield, secondary methods are also employed. This generally calls for pumping either water into the stratum below the pool or gas into the layer above, with the intention of flushing out the crude oil clinging to the reservoir rock. According to Earl Cook (1976), between 10 and 60 percent of the original *oil in place* in various reservoirs in the United States is recovered by these two methods, with the cumulative recovery average being about 31 percent. The thing to be noticed here is that for every barrel of crude oil produced, more than two will have to be left in the ground if their recovery is solely dependent on these methods at their present state of efficiency. It also appears that even low-cost oil (from $.25/bbl to $1.50/bbl) can become medium-cost oil (from $1.50/bbl to $4.50/bbl) when secondary recovery methods based on water flushing are used.

At this point it behooves us to distinguish between the terms *resources* and *reserves*. By resources we mean oil in place in the sense that we used this expression above, but at the same time stretching this concept to include both discovered and undiscovered deposits. Undiscovered resources are important since they include not only extensions of existing producing areas, but also new fields in new localities where, according to a consensus of geologists, oil should be found regardless of its present status in the exploration records. The east coast of Africa, including offshore areas, has some interesting geological traits, and there are abundant theories claiming that Egypt, which lies between the great oil-producing areas of the Middle East and North Africa, should be well supplied with petroleum. (And in fact, Egyptian output is expanding at an impressive rate.) As for reserves, these are resources which, given the prevailing state of science and technology, can be economically exploited. Figure 3-1 clarifies some of these ideas.

In conjunction with this figure, it should be mentioned that there are also tertiary recovery methods. (At present these are relatively unimportant, and there is an important body of opinion which claims that by any realistic analysis they are doomed to remain uneconomical.) Existing tertiary techni-

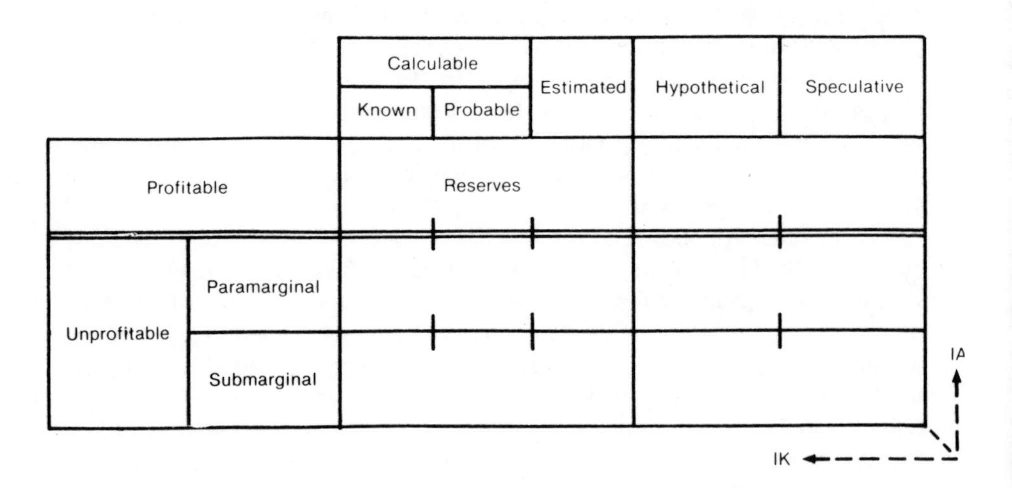

Source: U.S. Geological Service.

Reprinted by permission of the publisher from Ferdinand E. Banks, *Bauxite and Aluminum: An Introduction to the Economics of Nonfuel Minerals* (Lexington, Mass.: Lexington Books, D.C. Heath and Company). Copyright © 1979, D.C. Heath and Company.

Note: IA: Increased availability.
IK: Increased knowledge.

Figure 3-1. The Classification of Mineral Supplies

ques, for the most part, revolve on lowering the viscosity of oil in a reservoir so that it can flow more freely; and in practice this is brought about by heating the oil (by injecting steam into it) or by injecting chemicals directly into the reservoir. Because tertiary methods are used more liberally in the United States than elsewhere, the United States has a total recovery rate of 32 percent, as compared to an average world recovery rate of 30 percent; and it has been suggested that in time the recovery rate in the United States might reach 40 percent. This matter of the percentage of an oil deposit that is ultimately recoverable is extremely important, as the following example should indicate. In 1973 it was projected that Saudi Arabia was technically capable of producing 20 million barrels of oil per day (20 Mbbl/d) since at that time it was believed that the average recovery factor in that country could reach 47 percent; but on the basis of the performance of the main field, Ghawar, this figure has been revised downward to 33 percent. What this means is that the planning maximum of Saudi Arabian production must now be regarded as 12 to 13 Mbbl/d, while estimated recoverable reserves are reduced by 70,000 Mbbl, or more than the whole of Iran's proved reserves and equivalent to about three times the value of proved North Sea reserves.

Finally we close this section by making a few disparate remarks that are essential to rounding out the introduction. First we can note that in the United States the average daily production of an oil well in 1972 was 18.4 bbl; however, in the United States some famous wells have produced as much as 100,000 bbl/d. The deepest well in the United States today is 31,000 feet (ft), compared to 5,700 ft in 1908. It appears, though, that there is not a great deal to be gained by going below 17,000 ft, since at that depth geothermal heat and pressure transform oil into gas and, in addition, the weight of the earth reduces the pore space in potential reservoirs. It should also be noted that oil reservoirs are of widely varied sizes, and a large fraction of known reserves occur in so-called giant fields: 4 oil fields account for approximately 21 percent of the reserves of the noncommunist world; 29 oil fields account for 49 percent, and about 30,000 oil fields contain the remaining 51 percent. Despite rumors of the presence of new giant fields in Mexico and Venezuela, as of the end of 1979 it was possible to say that none had been certified in the past decade, although they are the prime target of explorers.

It is also essential for the reader to be aware that there are several concepts of capacity. The first is called *facility capacity,* and this refers to the total installed capacity of gas-oil separating plants, main-trunk pipelines, and oil-loading terminals. In March 1979, the facility capacity of Saudi Arabia was reputedly 12.8 Mbbl/d. *Maximum sustainable capacity* is the maximum production rate physically sustainable for at least six months, and usually this comes to 90 to 95 percent of facility capacity. For Saudi Arabia this came to 9.8 Mbbl/d at the date given. The last concept is *surge capacity,* and this is the maximum output that can be produced for a short period, such as a few weeks. As indicated by several production peaks, this level came to 10.5 Mbbl/d for Saudi Arabia in 1979.

The Reserve/Production Ratio

This topic is crucial to explaining why the world production of oil will decrease when there is still a large amount of recoverable oil in the ground. To understand this phenomenon, the reader must grasp the significance of the reserve/production ratio.

The general opinion today is that the amount of recoverable petroleum located in the continental crusts and offshore areas amounts to almost 2 trillion barrels (Tbbl). This figure emerged in 1977 at the World Energy Conference from a poll that included the seven largest U.S. oil companies and the U.S. Geological Survey, and as far as I know, it has not been seriously challenged since that date. At that time the accounting took the following form: since the discovery of oil 361 billion Barrels (Gbbl) has been produced, and an additional 652 Gbbl is listed as proved reserves. The other 997 Gbbl falls in the probable-reserves category in that it is not yet discovered, but most geologists feel certain that this oil will eventually become available.

Next we can emphasize that each oil field has a potential production rate that depends on the size of the field, its geology, and its facilities for lifting and transporting oil. In general, it is uneconomical to produce more than 10 percent of the recoverable oil in a field during a single year, since if this is done, the amount of oil that can eventually be recovered is reduced. The analogy that we can use to explain this phenomenon is a machine, or a vehicle, which can be run at very high speeds if we desire, although to do so over an extended period means that we greatly reduce its usable life. Actually a reserves-to-production (*R/P*) ratio of 10 is probably the absolute minimum that could be used for the world (or even a major producing country), since this would imply that all oil fields can simultaneously produce at the maximum rate, when the fact is that some fields will merely be under development even though they contribute to the aggregate of proved reserves. This reason usually suffices to explain why the critical *R/P* ratio is usually put at 15 instead of 10, although another reason is implied in chapter 4 when the effect of differential field size is examined.

A numerical example is now constructed to show how the *R/P* ratio influences production. Two cases are examined. In both the amount of initial reserves is taken as 220 units, but *intended* production is different for the two cases. In case 1, production is to be held at 10 units per period; in case 2, production is to grow at a rate of 10 percent a year. Thus *intended* production in the second case is 10, 11, 12.1, 13.3, . . . and so on. If we assume a critical R/P ratio is 15, figure 3-2 shows the time path of production for these two cases.

At this stage of the discussion it is essential for the reader to consider the following observations. In case 1, after the eighth year the minimum (critical) R/P ratio (= 15) takes over, and production falls below the intended level (= 10). Essentially the same thing is true when we have a growing production. As pointed out above, the intention is for production to

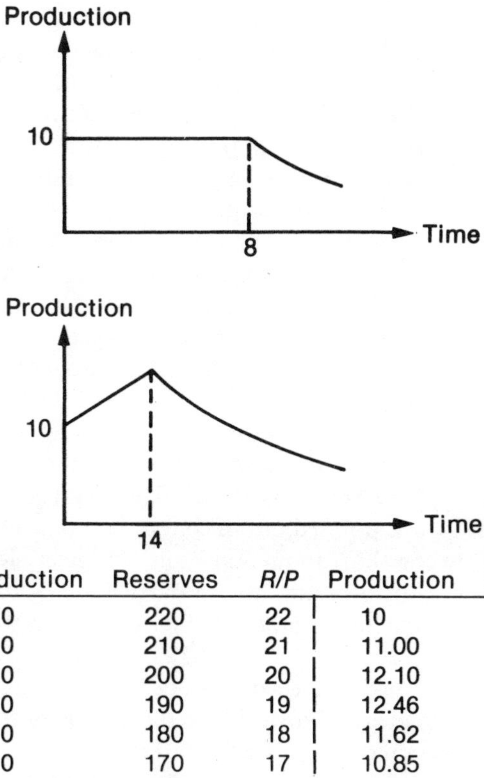

Year	Production	Reserves	R/P	Production	Reserves	R/P
1	10	220	22	10	220	22
2	10	210	21	11.00	210	19.09
3	10	200	20	12.10	199	16.40
4	10	190	19	12.46	187	15.0
5	10	180	18	11.62	174	15
6	10	170	17	10.85	163	15
7	10	160	16	10.13	152	15
8	10	150	15	9.45	142	15
9	9.33	140	15	8.82	132	15
10	8.77	131	15	8.23	124	15
11	8.16	122	15	7.68	115	15
12	7.58	113	15	7.17	108	15

Note: Some figures for reserves are rounded.

Figure 3-2. Production as a Function of Time, with a Critical R/P Ration of 15

grow as 10, 11, 12.1, 13.3, 14.6, . . . ; but after the third year the crucial *R/P* ratio is reached, and production starts to fall. Particularly notice that, in both cases, when production starts to fall, less than half of the reserves are used up (although this situation would be modified somewhat by an *R/P* ratio of 10 or 12). The thing to realize here, though, is that production falls because of profit-maximizing behavior by the owners of this oil, not because of any natural restrictions. In fact, in an emergency the original production plans could be carried out for a longer time, but the figure for reserves (220 units in the example) would have to be adjusted downward in the periods after the critical *R/P* ratio was passed, since this reserve figure is determined on the basis of the economic rate of recovery. Put another way, if it becomes necessary to exceed the economic rate of recovery for any substantial period, the oil deposit will be damaged in such a way that less than 220 units is eventually recoverable (in the asymtotic sense).

We may now ask just what this example tells us about the real world. To begin, it assures us that world oil production is also going to *peak,* and when this happens, there is going to be plenty of oil in the ground. The most important question remaining about this phenomenon is, when? Many optimists claim that peaking can be delayed until around the turn of the century, while pessimists insist that world oil production could begin to decline in the late 1980s (although, regardless of the date, it appears that world oil output at the time of peaking will be between 75 and 85 Mbbl/d). What, then, might delay the bad news? The most important thing would be a sharp decrease in the rate of growth of demand for petroleum or the discovery of a new giant field every few years. Without extensive rationing, the first seems unlikely since, as was pointed out in chapter 2, oil is an indispensable industrial ingredient in the short to medium run, and using less of it means that we will also use less of other inputs, in particular, labor. The already high level of unemployment in the industrial world suggests that this is not a healthy prospect.

Similarly, informed opinion seems to be discounting strongly the likelihood of another Ghawar or North Slope turning up in the near future, although in the next section it is suggested that if the intensity of exploration in the world as a whole were to match that of North America, the figure for global recoverable reserves might be altered drastically. Other possibilities are a decisive shift to new energy sources, to include some "oily" materials such as tar sands, and oil from shale and coal. Unless there is a breakthrough in solar energy or nuclear energy becomes more acceptable, this shift *will* take place; but some question must be raised as to whether it will come in time. As far as I can tell, the answer is no, principally for the following reason. In this kind of project, governments would have to cooperate with private industry in order to eliminate certain types of uncertainty; for example, these synthetic-fuel firms would have to be "insured" against the possibility of a conceivable (although unlikely) fall in energy prices that would make their operations unprofitable. At present it appears that both governments and the management of the firms that would be involved are amenable to this type of relation; but there is an important opposition that has managed to block the initiation of the necessary programs. At the core of this opposition are certain influential academics, as well as various politicians and journalists who mistakenly believe that either the present oil-market upheaval is a soap opera that would soon be over if the "free market" were allowed to prevail, or "soft" energy options such as solar devices and biomass will soon be able to supply us with all the fuel and electricity we need. Figure 3-3 summarizes the above discussion.

To close this phase of the exposition, let me say a bit about reserves. If some of those trillion barrels of oil scheduled to be found could be discovered a great deal earlier, then perhaps the industrial world would be able to delay, by another decade or so, making the huge investments eventually required in nonoil energy sources. But as things stand, the rate of discovery of oil may have dropped onto a lower trend. In broaching this problem, it is essential to realize that the manner in which discoveries are listed is important. The figures showing total reserves can be examined on the basis of year end to year end (where, if desired, these figures can be smoothed by using three- or five-year averages); or, more relevantly, reserve figures can be backdated to the year in which the field was discovered. This works as follows. Assume that a field discovered in 1970 contains 1 Gbbl of

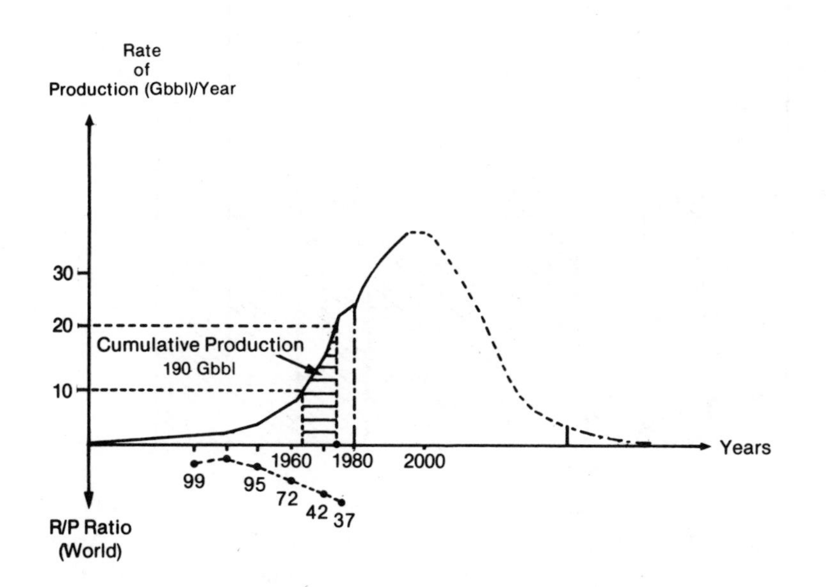

Figure 3-3. Cumulative World Oil Production and the World R/P Ratio

recoverable oil and in 1975 this figure is increased to 2 Gbbl. On the basis of year-end accounting, the extra billion barrels would be attributed to 1975, while backdating would assign it to the 1970 discovery. As shown in figure 3-4, backdating breaks up large reserve increases which may take place in just a few years into components that are assigned to a larger number of years.

Although drilling activity may be setting a new record in the United States over the past few years, with the number of wells being drilled per year doubling from the 1973 figure of 25,000 to 50,000, little new oil is being found. (In addition, the United States recently has been the site of several spectacular exploration failures—Baltimore Canyon off New Jersey; Destin Dome, which is offshore and adjacent to the Florida Panhandle; and Cook Island and the Gulf of Alaska. Lease rights alone for these

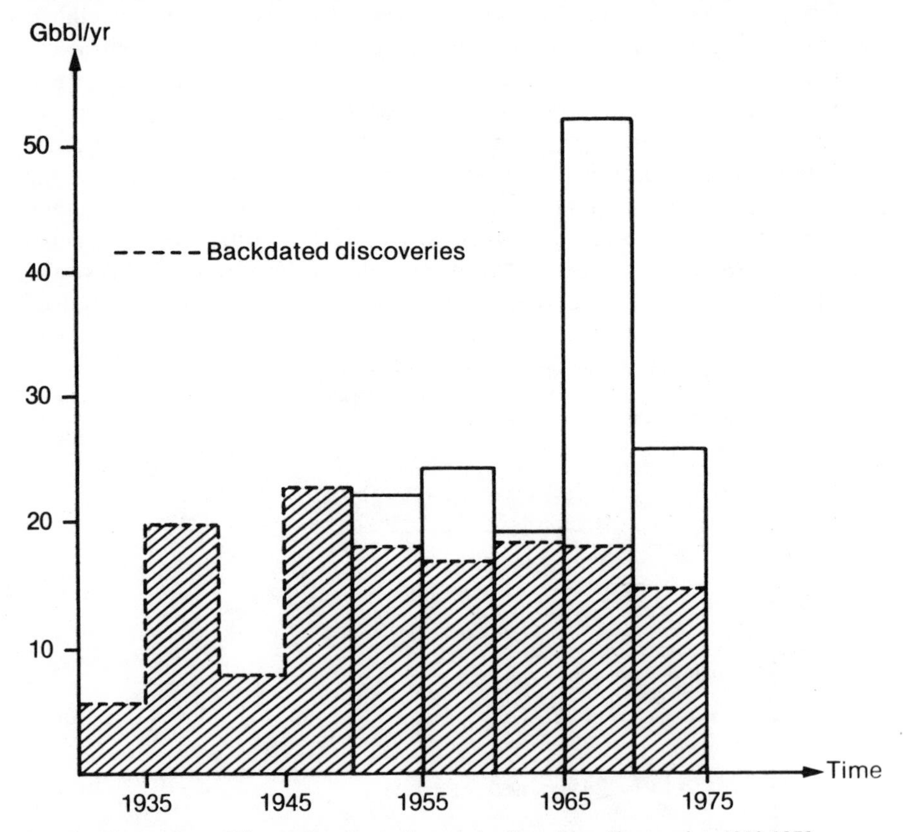

Source: Figures from Oil and Gas Journal, and the Petroleum Economist, 1950-1978.

Figure 3-4. World Oil Discoveries as Five-Year Averages of the Annual Increase in Proved Reserves (Solid Line) and Backdated to the Year in Which the Oil Field Was Originally Discovered

miscarriages came to billions of dollars.) In the world as a whole, newly located reserves amounted to an average of approximately 18 billion barrels per year (Gbbl/yr) between 1950 and 1970; but since 1970 the figure has been 15 Gbbl/yr. As things now stand, about 20 Gbbl/year should be found in order to keep reserves from reaching a dangerous low, given the rate at which consumption may have to grow in order to permit the economies of the industrial world to maintain a decent rate of growth for the rest of this century. But in 1976-1977 oil discoveries actually sunk well under 15 Gbbl/yr.

One country that has not been mentioned thus far is the Soviet Union, which is the largest producer of oil in the world and exported at least 1 Mbbl/d of its 11.7-Mbbl/d production in 1979 to the noncommunist world. Some theories are now being advanced that production in the Soviet Union will soon start falling and that by the last half of the 1980s that country will be a new importer (rather than exporter) of petroleum. Specifically it is being said that the Soviets have already found much of the easily locatable oil in their country, and the development of new reserves will therefore require operations in Siberia and the far north, both of which are high-cost areas. This may or may not be true, since the Soviet Union also has a major nuclear-energy program underway and has access to considerable domestic supplies of coal and gas. Given the comparatively small amount of exploration that has taken place in Russia in relation to its size, as well as the historical rate at which production has been expanding, it would be very surprising if their output of petroleum suddenly reversed direction unless it were the explicit intent of the Soviet government.

Drilling and Exploration

In this section we expand on the perfunctory remarks about drilling and exploration made in the previous section, but first we talk about the exploitation decision as it might be viewed by an oil company or a government.

If we make the realistic assumption that oil is searched for and lifted in order to make money, then it could be argued that the way to maximize profits from this operation is to maximize the monetary benefits from producing oil relative to the monetary costs involved. The reader looking for a brief survey of cost/benefit analysis as it applies to this topic can find one in Banks (1979b), but roughly what we do is as follows. First we list all benefits in the form of estimated receipts from selling oil on one side of a ledger; then we put the costs incurred on the other side. These benefits and costs involve different points of time, and so they should be put on a common basis (usually by discounting them to the present year). Then if benefits (or, discounted revenues) outweigh costs (discounted) *and* sufficient importance is attached to how uncertainty enters calculations of this type, the project is judged viable.

Some of the costs that must be considered are seen in figure 3-5, as well as the expected flow of receipts. These costs include such things as exploratory costs (for example, seismographic examinations and exploratory drilling), which can be very high under certain circumstances. Later, if the field is worth exploiting, there are production costs which include items such as the capital cost of equipment and the wages of employees. It is worth calling attention to the comparatively low outlays that are often experienced here. In the United States the operating cost per well in 1976 was about $9,000 on average, which meant a *per barrel* operating cost of less than $.07. Other highly relevant costs are taxes, social charges of various types, and perhaps a considerable amount of infrastructure in the form of roads. The time factor is also important. It takes years from the first seismographic examination until maximum production is reached in most oil fields.

One thing that must be remembered here is that most of the costs and benefits referred to above are events of the future. This means that in a diagram such as figure 3-5, sketched at the time leases are purchased and hopes are high, *actual* costs and benefits are unknown. For instance, in the drilling failures referred to earlier, there were no benefits, and a large part of this kind of diagram would have been irrelevant.

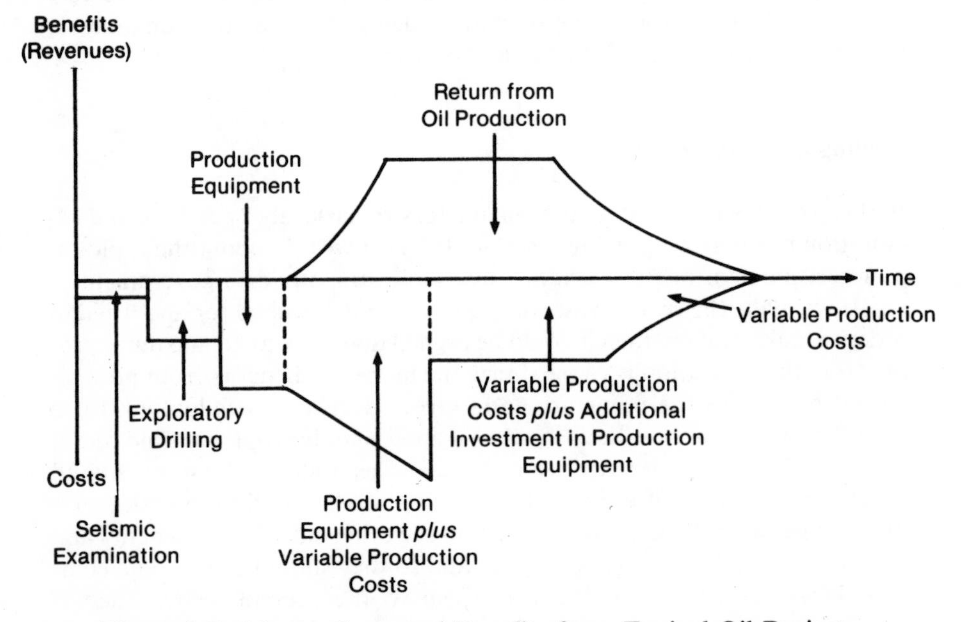

Figure 3-5. Money Costs and Benefits for a Typical Oil Project

This short section closes with a caveat referred to earlier. The general belief of the major oil firms is that only a few of the unexplored or recently explored areas show signs of having the geological characteristics underlying supergiant producing fields. Moreover, because of advanced technical equipment and increased geological knowledge, it is becoming progressively easier to determine whether a prospective site is worth drilling. In the United States this has meant that locating a "significant" field (which is one containing at least 1 Mbbl of oil) has required only forty-eight wildcat wells in 1978 versus seventy in 1961. Thus, given the tens of thousands of exploratory wells that have been drilled over vast areas where, according to geological science, oil *could* be found, and given the fact that a supergiant field has not been uncovered anywhere in the world since 1968, a reasonable suspicion might be that we have accounted for just about all these Eldorados.

Yet it must be remembered that of the more than 3 million oil wells drilled in the last 120 years or so, 2.4 million were drilled in the lower forty-eight states of the United States. Of the exploratory (as compared to development) wells drilled before 1976, 95.4 percent were drilled in the industrial world. Africa, all Asia, and Latin America are just about untouched. Even Australia, Canada, and the Soviet Union are comparatively unexplored. For instance, in Australia only fifty-three exploratory wells were drilled in 1978. In the Middle East there is very little exploration, since on the basis of its production a few years ago, only Iran should be concerned with its reserve position. On the basis of rather sketchy information it appears that since 1974 Saudi Arabia has averaged only ten exploratory wells per year; Iraq, only one; and the entire Middle East, only ninety-five per year. As far as I know, only Libya and Algeria are aggressively seeking oil company help with exploration. Libya has informally adopted the slogan "no oil supply without exploration," while Algeria (which only has a production of 1 Mbbl/d and reserves of 8.4 Bbbl) imposes a surcharge of $3/bbl on current contract sales, but refunds this amount if the purchasing oil company can offer an acceptable exploration program.

Thus the last word about the availability of giant fields cannot be be said until some further tens of thousands of exploratory wells have been scattered around places like Madagascar, the Chaco in Bolivia, and some of those other environments which oil producers classify as moderately harsh (such as Central Africa) or extremely harsh (such as Artic areas or places where drilling must take place in water as deep as 6,000 feet). These activities are going to cost billions; and whether private oil companies will continue to be willing to make investments of this particular nature over a long period if, in addition to encountering large numbers of dry holes, they have to support the kind of public hostility that has fallen on them lately, is quite dubious.

***User Cost and Backstop Technologies**

The concept of a *user cost* (or *scarcity rent* or *scarcity royalty*) has become a popular topic for economic theorists since the onset of the energy crisis. It is basically a very simple idea which was explained more than adequately by Hotelling (1931) almost fifty years ago in the context of one of the first "modern" expositions of exhaustible resources. Davidson (1963) has treated user-cost theory in connection with some problems related to the U.S. crude-oil industry. According to him, its purpose is to point out that since it is true that the more oil that is used today, the less will be available tomorrow, the price (or marginal revenue) of oil must cover not only marginal operating costs but also the present value of marginal profits given up by producing now instead of later.

Thus, even if a resource can be extracted at a constant (marginal) cost until it is used up, its owner would not be committing a sin against basic economic theory by refusing to sell any unless its price were higher than marginal cost. This is so because if that resource *were* in the process of becoming scarce, an *efficient* market on which the resource was being sold would establish a price above the marginal cost; otherwise, knowing that the resource was becoming scarce, its owners would leave it in the ground and extract it later when its price did rise. Just how much above marginal cost the price would have to be depends on the degree of scarcity. In particular, it should be appreciated that if the resource *were* manifestly scarce and for some reason the price were listed as being equal to the marginal cost—which, as we remember from elementary economic theory, is the usual profit-maximizing criterion—then producers would withhold it from the market. Under these circumstances, if the resource were an essential input (such as oil), then the necessary price rise would be evoked by their actions. In fact, the price might rise so rapidly as to lead to a self-perpetuating price acceleration which, although eventually checked, causes large amounts of the resource to be sold at unnecessarily high prices. We may have been witnessing some of this kind of behavior in the last part of 1979 when, on the Rotterdam spot market, oil that very likely could be bought for much lower prices in the middle of 1980 (and which would not be needed until that time) sold for double the OPEC price because of consumer fears about the future availability of this commodity. Moreover, the movement of these spot prices undoubtedly influenced the setting of posted (OPEC) prices, causing them to be higher than they would have been otherwise; and probably influenced future OPEC policy as to output.

The same thing could happen in the other direction. When the huge oil strike was made in east Texas in 1931, oil producers, expecting the worst, flooded the market with oil. The price of U.S. crude petroleum dropped rapidly from $1/bbl to $.10/bbl, and even at this price the market was not cleared. The upshot of all this is that price-marginal cost gaps are *not* sufficient evidence of resource misallocation in a market for an exhaustible resource, and it is fully appropriate that, in a competitive market of the textbook variety, with approaching physical scarcity, price can become very large even if extraction costs are near zero at the margin. As is argued in chapter 4, it is not all self-evident that before the 1979 price rises the OPEC-established world oil price was excessively out of line with that which would have come about had oil prices been completely determined by so-called free-market forces. Readers desiring brief and simple algebraic explanations of these phenomena are referred to Solow (1974), Siebert (1979), and Fisher (1979); a comprehensive but elementary survey of the same topic has been made by Lecomber (1979); while pioneer work on this topic is being done by John Hartwick, Michael Hoel, and Murray Kemp.

Next we look at the concept of a backstop technology. Nordhaus (1974), in a brilliant but overoptimistic article, defines a *backstop technology* as a substitute process with an infinite resource base. Actually, a very large resource base is adequate for our purposes. A backstop technology for the oil industry would be one that could provide an oil-like substance for many decades. If there are several, such as oil from coal and oil from tar sands, then, strictly speaking, the backstop technology is the one that is least expensive. One of the things a backstop technology is supposed to do is determine the highest price that can be charged for a particular resource. At present it is said that if the price of crude petroleum were to exceed $40 and there were some assurances that it would remain there, then it would be profitable to construct plants capable of producing a large amount of synthetic oil from coal or to accelerate the mining of oil from the huge deposits of oil in tar sands located in such places as Venezuela, Canada, Russia, and (to a lesser extent) the United States.

But if we look more closely at the situation in the real world, where the price of oil almost doubled in one year, we see that a $40/bbl oil price might be just around the corner, particularly if there is more political trouble in the Middle East; however, the facilities for producing nonconventional oils are being constructed on only a minor scale. Thus some indication that the price of oil is en route to $50/bbl may be required before the transition from crude petroleum to oil produced from coal or tar sands assumes serious proportions. The so-called oil club of the International Energy Agency did manage to reach some kind of agreement when the price of oil increased by a factor of four to $11.65 in 1973-1974 and provisionally they accepted a "floor" price of oil of $8. (What this seems to have involved was taxing oil that could be imported for less than $8 in such a manner that the price to purchasers was $8.) Some resistance would undoubtedly be met, though, to the establishment of a floor price in the vicinity of $40; and yet if synthetic or heavy oils were to be successfully ushered in, an arrangement of this type might have to be imposed in order to prevent the new oils from being periodically undercut by producers of conventional oil. After all, in ten or twenty years there will still be producers in the Middle East and elsewhere who can supply copious amounts of crude oil at less than $5/bbl and still make a profit.

We conclude this chapter with a simple example designed to raise some of the issues associated with a backstop technology. To begin, let us assume that we have a tropical island whose population consists of fifty-five people called islanders who need to work only one hour per year. However, this is quite adequate to provide them with sustenance and leaves them with plenty of leisure time to devote to their two favorite sports, singing and dancing the light fandango, which together are called "forskning." Let us also assume that this society is a stagnant one, islanders live forever, and reproduction is unnecessary. All work is done on one day of the year, January 1, when this society produces 100 units of product called *sill* whose inputs are rotten fish that wash up on the beaches (and are therefore free), oil (for rinsing the fish), and labor. The oil comes from a pool near the beach which, at the time we begin our story (just before the start of work on January 1, 1980), contains 50 barrels (bbl). This year, as every year in the past, during the hour of work five islanders will withdraw 10 bbl of oil from this pool since, for technical reasons, each unit of sill produced requires 0.1 bbl of oil as a rinse.

At some time in the past, the islander "government" printed 55 dollar bills, and this suffices to buy the output of sill and pay the islanders. Mechanically, this procedure goes as follows. When this idyllic society was created, one of the islanders was designated boss. He has the receipts ($55) from the previous year's sale of sill in his pocket when work begins at 11 o'clock on January 1. At 12 o'clock, when the work for that year is over, he pays all the workers (to include himself) $1, and here we should note that this means both those islanders producing sill *and* those producing oil. At 12:01, after carefully counting their pay, the islanders buy, for $55, 80 of the 100 units of sill produced. The boss now has the gross national product of the country ($55) back in his pocket, and another year's work is over. Enough food has been produced to last these men another year, and they can return to their valuable "forskning."

But now we have a problem. The islanders produced 100 units of sill, but they consume only 80 units. What is the explanation for this odd state of affairs? If we assume a perfectly competitive economy of the textbook variety, there can be only one explanation in light of our previous discussion: If 10 bbl/yr is being taken out of the oil pool and it contained 50 bbl at the start, then it will only last five years. After five years this domestic oil will have to be replaced by foreign oil or by something equivalent to oil; and what is happening is that the 20 units of sill not going to the islanders to be consumed this year are being used, in some sense, to pay for this new source of oil that will be required in five years.

Let us inspect this situation a bit more closely, and to keep the exercise simple and to the point, let us make the following assumptions. First, no foreign oil is available. Second (and here we introduce the concept of a backstop technology), it is possible to import from Japan a nondepreciable machine which, together with five islander operators, can produce the oil required each year from the infinite supply of free coral on the island.

The 20 units of sill per year that are not consumed are therefore being exported to Japan in order to pay for this machine. We also assume that the first shipments are made five years before the machine is required (on January 1, 1985) and that payment for the sill is in yen at a price of one yen per sill. These yen are then put in a savings account in a Japanese bank where the interest rate is 10 percent, and in five years' time the accumulated savings, together with interest, suffice to buy the machine. The deposits and interest accumulated at the Japanese bank are shown in figure 3-6. As for the cost of the machine, on the basis of the above assumptions, this must be $20(1 + 0.1) + 20(1 + 0.1)^2 + 20(1 + 0.1)^3 + 20(1 + 0.1)^4 + 20(1 + 0.1)^5 = 134$ yen.

We could look at this calculation from another angle. The Japanese manufacturers of this component of the backstop technology quote a price of 134 yen for the machine on the date they deliver it to the island, which is January 1, 1985. If the islanders start saving for it five years in advance, then they must save 20 sill per year whereas, as is shown below, if they had started saving ten years or more in advance, then considerably smaller year-

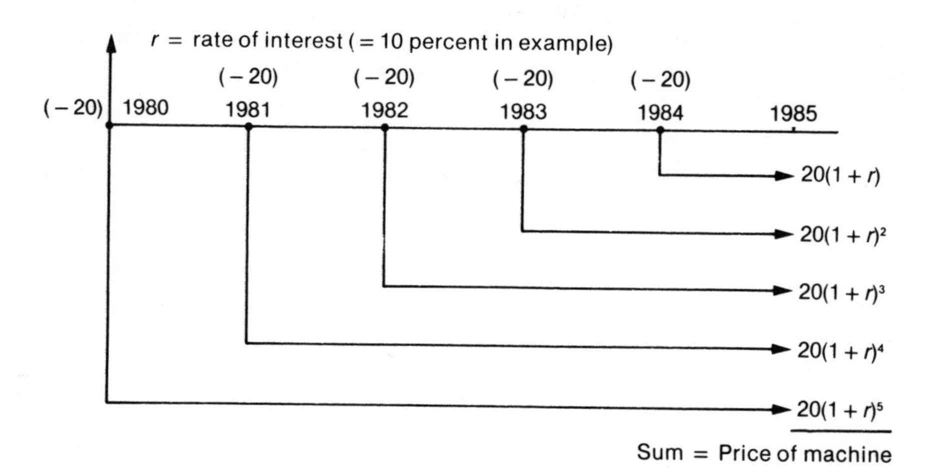

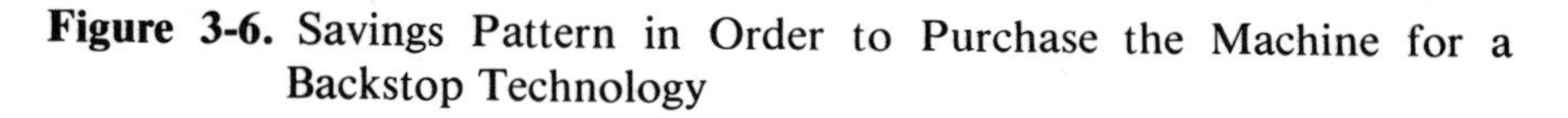

Figure 3-6. Savings Pattern in Order to Purchase the Machine for a Backstop Technology

ly savings would have been necessary. There is still another way to view this matter. In five years, with a market rate of interest of 10 percent, 134 yen is equal to $134/(1 + 0.10)^5 = 83.2$ yen now. Thus 83.2 is the total scarcity royalty (or user cost) of an indefinite yearly consumption of 10 bbl of oil. Notice the phrasing here and the reference to both quantity and time. Also notice that on the basis of our previous calculation this 83.2 yen can be divided into five payments of 20 yen per year over a period of five years, if we assume that the interest rate is 10 percent.

Similarly, had the islanders started saving ten years before their oil ran out, the total scarcity royalty would have been 51.6 yen; and this could have been divided up into ten payments of 7.65 sills per year (again assuming a sill price of one yen per sill and an interest rate of 10 percent). At the extreme, had this coming shortage been recognized one-hundred years before it took place and had the islanders begun to save at that time, the total royalty would have been 0.00972 yen. The simple algebra on which this example is based is shown first in appendix 3A, then in 4A. But one more observation is relevant now. The decision to begin saving for the oil-making machine five years before it was needed is only one of a large number of similar decisions that could have been made: saving for this machine could have begun ten years before domestic oil ran out, or twenty, or a thousand. Deciding which arrangement is suitable depends on the preferences of the islanders in the sense that somehow, out of myriad possibilities, one was chosen. But, by the same token, in terms of this illustration, no prior saving at all had to take place: On the day of delivery 134 yen could have been borrowed and, with an interest rate of 10 percent, repaid in an infinite string of yearly installments, with each payment being equal to $134 \times 0.10 = 13.4$ yen (= 13.4 sills if the price of sill continues to be 1 yen per sill). This, however, is a textbook ploy. In the real world it is often much easier to save than to borrow.

Appendix 3A: Backstop Technologies and a Comment on Investment Theory

The simple algebra underlying the backstop-technology example is discussed here. To begin, we can show our model in the form of 2×2 input-output table with one primary resource—labor (b_o and b_s) working in the oil and sill sectors, respectively:

	Oil	Sill	F	T			Oil	Sill	F
Oil	a_{oo}	a_{os}	F_o	T_o	=	Oil	0	$\frac{1}{10}$	0
Sill	a_{so}	a_{ss}	F_s	T_s		Sill	0	0	100
Labor	b_o	b_s				Labor	$\frac{1}{2}$	$\frac{1}{2}$	

With the exception of b, the notation here is the same as in the second section of appendix 2A-2. The following simple algebra applies to this scheme:

$$a_{oo}F_o + a_{os}F_s + F_o = T_o$$

$$a_{so}F_o + a_{ss}F_s + F_s = T_s$$

Using the simple coefficients that are supplied, we get $T_o = 10$ and $T_s = 100$. Where total labor (B) requirements are concerned, we have $B = B_o + B_s = b_oT_o + b_sT_s = 0.5(10) + 0.5(100) = 55$ labor-hours which, providentially enough, is equal to the number available. Now we know that gross national product (GNP) is equal in this simple arrangement to payment to primary factors which, in turn, is equal to the value of output. Thus we can write, where w is the wage,

$$\text{GNP} = w(B_o + B_s) = P_oF_o + P_sF_s$$

At this point, a *numeraire* is required, and so we set $w = \$1$. Then we have $1 \times 55 = 100\ P_s$, or $P_s = 0.55$. What about the price of oil? One unit of oil requires 0.5 labor-hour, and since 1 labor-hour costs \$1, $P_o = \$.50$. Of the sill price, however, only \$.05 can be attributed to oil, since 1 unit of sill requires 0.1 unit of oil. Here our price equation is $P_s = 0.5\ w + 0.1\ P_o = 0.55$, which corroborates the previous calculation for P_s.

Now let us note the irregularity introduced into our problem by the presence of the exhaustible supply of oil and the need to import (in five years) a Japanese machine to be used in a backstop technology. This machine will be paid for by exporting some of the yearly output of sill; and at a price of 1 yen per sill it was decided to export 20 units of sill per year. Putting this money in a Japanese bank (where it will draw 10 percent interest) means that in five years a machine will be paid for which, together with 5 units of labor and the infinite amount of free coral on the island, will produce the oil needed to produce the 100 units of sill that inslanders have decreed as necessary for their way of life.

Given *this* situation, where 80 sill is purchased with the national income of \$55, instead of 100 sill, the price of 1 unit of sill becomes \$0.6875. Where, now, should this adjustment take place? Since the oil sector takes charge of this community's oil supply, the unit price of oil should be increased above the \$.50 obtained earlier, just how much can be obtained from the price equation for sill. Thus $0.6875 = \frac{1}{2} w + \frac{1}{10} P_o = 0.5 + \frac{1}{10} P_o$, or $P_o = 1.875$. The user cost or scarcity rent per unit of oil is evidently $1.875 - 0.5 = \$1.375$ per unit of oil, and this remains the case until the backstop technology is installed.

Now we present a simple exercise having to do with the development cost of a resource such as an oil field. This type of analysis has its roots in elementary capital theory and employs the same type of reasoning that would be appropriate for any asset. To begin, we can examine two types of investment situations. (See figure 3A-1.)

In figure 3A-1*a* consumption is at $\overline{C}$ to begin (at time t_0). In the period from t_0 to t_1, it is reduced to $\overline{C} - x$ in order to realize $\overline{C} + y$ in the period from t_2 to t_3. According to the usual tenets of capital theory (and common sense), we can define the rate of return ϱ on this project in the following manner:

$$\frac{\Delta C_1}{\Delta C_0} = \frac{y}{x} = \frac{\text{return}}{\text{cost}} = \frac{\text{return} - \text{cost}}{\text{cost}} + 1 = \varrho + 1 \qquad (3A.1)$$

$$\varrho = \left| \frac{y}{x} \right| - 1 \qquad (3A.2)$$

We would expect $y > x$ since otherwise there would be no point in abstaining from x (although, admittedly, there are situations when this is not true).

In the situation shown in the first part of figure 3A-1*b*, h is given up over the interval from t_0 to t_1 and this, gives rise to a return z over an infinite period. (This would be equivalent to the purchase of a bond known as a perpetuity.) Now we have, with p being the price of this commodity and r a (constant) discount rate,

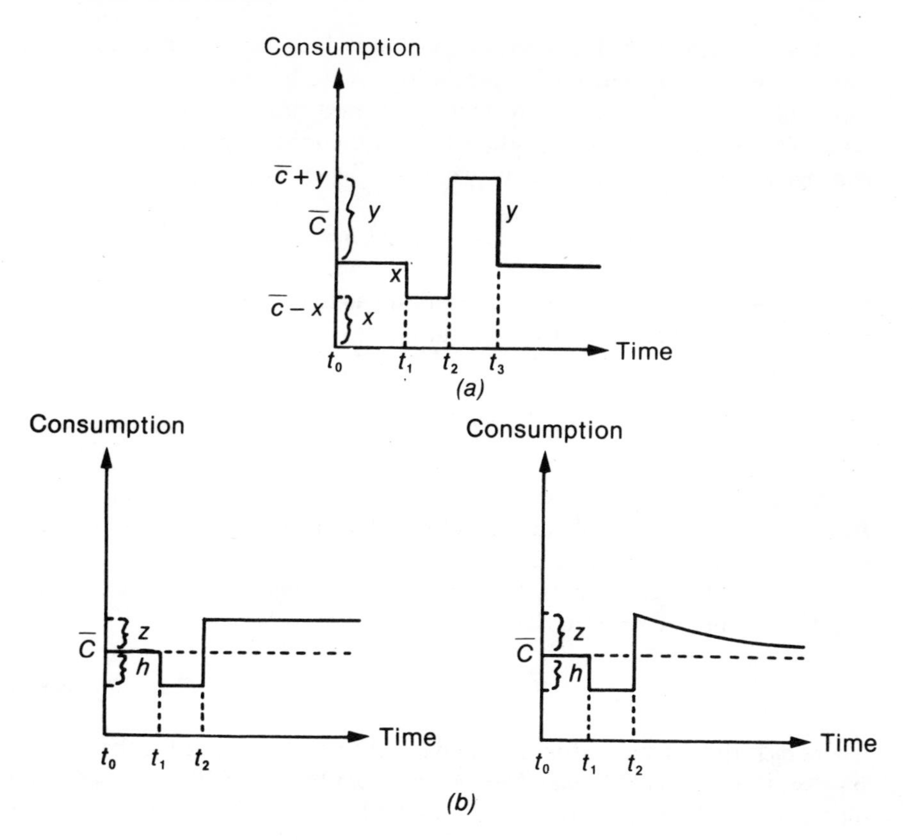

Figure 3A-1. Investment and Subsequent Patterns of Consumption

$$ph = \int_{o}^{\infty} pze^{-rt}\, dt = \frac{pz}{r} \tag{3A.3}$$

Thus

$$z = rh \tag{3A.4}$$

It should also be clear that an analogy to equation 3A.2 could have been obtained by using simple discounting on the arrangement in figure 3A-1A. Thus:

$$px = \frac{py}{1 + r}$$

or

$$r = \frac{y}{x} - 1$$

The problem with this exposition is that it does not take into consideration the very real problem of depreciation. If we have machinery or structures, they can be expected to depreciate over time. If this depreciation takes place exponentially (or radioactively), then the yield of these facilities is progressively diminished. Thus we would have, as in figure 3A-1b:

$$ph = \int_0^\infty pze^{-rt}e^{-at}dt \qquad a>0 \tag{3A.5}$$

Observe here that radioactive depreciation has the effect of driving the yield z asymptotically toward zero. Integrating this, we get

$$r = \frac{z}{h} - a \qquad \text{or} \qquad h = \frac{z}{r+a} \tag{3A.6}$$

As mentioned, the effective yield has been reduced by the depreciation. Another interpretation of these results is that in the presence of depreciation, the price of the facilities must be reduced [from z/r to $z/(r + a)$].

Now we can go to an oil field. Here recoverable reserves R can be defined as cumulative output. Thus

$$\overline{R} = \int_0^b q_0e^{-at}dt \tag{3A.7}$$

where q_0 is the current output rate and a is the rate of decline for the field. If $b \to \infty$, then we get $R = q_0a$. Next we can recall that as with any asset, if an oil field is to be developed, the discounted value of its net receipts must equal the cost of investment. Then we get, with I being the cost of investment and p the income per barrel

$$I = \int_0^b pq(t)e^{-rt}\,dt \tag{3A.8}$$

With $q(t) = q_0e^{-at}$ we get

$$I = \int_0^b pq_0e^{-at}e^{-rt}\,dt \tag{3A.9}$$

or

$$p_{\min} = \frac{I(a + r)}{q_0(1 - e^{-(a+r)b})}$$

(Notice that if $b \to \infty$, we get a direct analogy of our previous results.) The income per barrel required to bring about 1-unit increase in output is $p_{\min}$, with the cost of investment given and the pattern of output known. For more detail on these and similar matters, see Bradley (1967) and Adelman (1974).

The Price of Oil

Making a sensible statement about the future price of oil is a much simpler matter today than it was just a few months ago. Not only is the price of oil going to rise, but it is going to rise at an uncomfortable rate, just as it recently finished doing. Basically what has happened is that the OPEC directorate and their economists have learned that, contrary to the expert advice provided by a number of distinguished economists in the major oil-importing countries, there is only an indirect conflict between their short-run revenue requirements, or desires, and the maximization of long-run financial benefits from the sale of oil (where the term *indirect* obviously carries a heavy political connotation). From a strictly economic point of view, the major oil-producing countries are, at present at least, free to adopt a radical position on both price and output, although, as is made clear later, the visible depletion of cheap, conventional oil is almost as important in determining the price of oil as the monopolistic powers of OPEC.

In this chapter I expand this theme, employing essentially a verbal analysis. I am particularly interested in building on the exposition in the previous chapters to say something about the supply of Saudi Arabian oil, which is a crucial entry where OPEC production is concerned; and in this context I want to illustrate an interesting supply phenomenon that can arise because of differences in the size of oil fields in a major oil-producing country. I also extend the discussion of backstop technologies begun in chapter 3 to the point where a crude calculation can be made of the appropriate (competitive) price of oil in terms of the reserve situation as we know it today relative to some of the more prestigious projections of consumption. Of course, both these factors could change for one reason or another (such as major new discoveries or drastic cutbacks in oil consumption), but in the light of information available when this book was written, I will argue that before the December 1979 price rises, the actual price of oil was only slightly out of line with its theoretical value.

There is no point in denying that many people will disagree with this assessment. Professor Peter Odell, for instance, has consistently distinguished himself by insisting that almost all present estimates of petroleum reserves are too low and that the 2 trillion barrels of recoverable oil estimate given earlier will probably be doubled and may eventually be quadrupled. Now so-called mainstream economics, despite its many shortcomings, clearly distinguishes between gambling and making rational decisions based on all available information. Thus the issue is not whether the true oil reserves in the world are those claimed by Exxon, the International Energy Agency or by Odell, but whether the industrial world should prepare itself for a serious shortage of oil in the *immediate* future, which (regardless of the actual amount of oil in the crust of the earth) could mean unemployment rate of up to 10 percent in some countries and social turmoil of the type experienced during the 1930s and 1960s.

I conclude the chapter with a short discussion of the importance of inventories. What the reader should realize here is that although almost all the basic textbooks in economic theory bypass this topic, it is crucial for understanding the pricing of all the important minerals (including oil); and, as pointed out by Banks (1977, 1979b), the reader choosing to ignore it can rest assured that he or she will never have more than a superficial insight into why primary-commodity prices behave so peculiarly.

The spot (free-market) price of oil today is $10 to $15 more than the prices *posted* by OPEC, even though enough oil is being shipped to the industrial countries to ensure that current industrial and consumer activities can be carried out. The problem is uncertainty about future supply and demand; and thus it is natural for consumers of oil, and speculators, to hold inventories of oil (or *stocks,* as they are also called) amounting to a certain fraction of current consumption. It appears that this fraction is in the process of being adjusted upward, and as a result of buying for inventories, the spot price of oil is being supported at a level which is unnaturally high given the demand for oil to be used in current production or consumption activities. (A recent Exxon study, focusing on what it calls "usable stocks," claims that on October 1, 1979, these were 660 Mbbl, or a 16-day supply, in the United States, Europe, and Japan, as compared to 550 Mbbl a year earlier. Looked at another way, stocks were 11 percent above the historic average in the major consuming regions on October 1.) Furthermore, when the spot price stays for a long period above the official OPEC price, it has a tendency to "feed back" and influence these prices. Specifically, it causes them to be summarily raised from time to time by some OPEC sellers, promulgating an illusion of immediate scarcity, which in turn increases the attractiveness of a strategy that calls for keeping oil as scarce as possible. Thus in a period when oil prices have escalated at a rate exceeded only in 1973-1974, some oil-producing countries have announced plans for decreasing production next year, while in others all net investment in oil-producing facilities has been canceled or drastically scaled down.

It is also interesting to point out here that, from a theoretical point of view, it seems likely that even if overflowing inventories in combination with an economic slowdown in the main oil-importing countries lead to a sharp fall in the spot price of oil, and possibly many OPEC sellers offering discounts or rebates on their posted prices, then the possibilities are that in this "game" between oil consumers and producers, OPEC can still be declared a net winner over the price cycle that began early in 1979. This is so because their total gain on sales made via the spot market, together with revenues from the various surcharges added to the official oil price during 1979, should exceed the aggregate of discounts that will have to be offered later in the event of an oil glut.

A Price of Oil

We can begin the main part of our discussion with a few words about an important paper by William Nordhaus (1973), which appeared in the wake of the 1973-1974 oil price rises. Noteworthy in ambition and in some of its technical details, but flawed in its interpretation of the realities of the evolving energy situation, this document argued that it was a mistake for the OPEC countries to elevate the price of oil in such a brusque manner. According to Nordhaus, a 400 percent rise in oil prices would drastically alter the profitability situation for materials that were competitive with oil. Coal, oil from shale and tar sands, nuclear devices, and so on would now be able to increase their share of the energy market; and eventually the global supply picture for energy would be altered in such a way as to place a downward pressure on the price of oil. In turn, this would lead to the discounted profit stream of oil producers being considerably less than it would have been with a more moderate price rise.

As attractive as this reasoning may appear, and despite the fact that Nordhaus was able to create an impressive analytical apparatus that ostensibly proved the folly of huge oil price rises, to many it was clear from the beginning that arguments and models of the type Nordhaus employed were useful only if the right people believed in them—and by the *right people* I do *not* mean economists in Gothenburg. Given the structure of the world oil market in the early 1970s, the demand for oil, and the rate of growth of demand, the only way its price could have been restrained would have been for producers to continue to raise output at an even more rapid pace than was practiced before 1973. Not only would doing this have required huge investments in new capacity, infrastructure, and so on, but also the unbridled consumption of oil at the traditional rate may have resulted in the peaking of world oil production before the end of the 1980s. As a result, many of the oil-producing countries would face an irreversible decline in the physical output of their most valuable and, in some cases, only natural resource. They would also have been put in the unenviable position of having to watch most of the investments and some of the infrastructure referred to above become superfluous. Equally disconcerting, consumers in the main industrial countries would find themselves with a stock of durables whose efficient utilization was dependent on an energy resource whose price, because of its increasing scarcity relative to demand, could begin to escalate in giant increments.

It is also important at this time to recognize that, depending on their aspirations in regard to future economic development, it might be advisable for many of the OPEC countries to be in possession of a sizable quantity of their oil reserves in a decade or so, even if there were a decisive technical breakthrough in the production of some source of cheap energy which led to a decline in the price of oil. This is so because perhaps in the not too distant future some of these countries can manage to accumulate the trained workforce which will allow them to use this oil as a feedstock in local refineries, petrochemical installations, and various other energy-intensive industrial activities (such as the smelting and refining of nonfuel minerals). See, for example, the reference to the Saudi Arabian petrochemical projects in chapter 7. Here we can acknowledge that while industrial efforts in many OPEC countries have not always been impressive, the disadvantages of placing OPEC revenues in financial instruments could easily become so great that most of these countries would have no choice but to continue to invest in physical assets (and concentrate on the development of their educational systems); and even if the next few years sees a succession of technological masterpieces hatched by solar and nuclear scientists, it will still take such a long time to introduce this technology on a large scale that anyone in possession of oil after the turn of the century will probably have a profitable outlet for that product. Economic theorists will also appreciate that the value (price) of a commodity such as oil can be understood only in terms of the consumer goods or investment opportunities that it will make available in the future, as well as the price that these will command in the market. As things now stand, these future opportunities could be considerable.

In chapter 3 it was pointed out that as the price of oil climbs relative to the price of other energy resources, it becomes profitable to introduce larger amounts of these alternative materials. We saw in 1973-1974 a kind of reversal of the declining enthusiasm for coal in the United States, and it has been said that if the price of oil reaches such and such a level, then the production of oil from coal will be possible in coal-rich countries such as Britain, Germany, and the United States. The technology for producing this "synthetic oil" was called a backstop technology (although, as pointed out, it lacked some of the qualifications for being a backstop technology in the sense in which Nordhaus defined this term); and in the context of this discussion, the same terminology would apply to the equipment used to produce oil from tar sands, since the cost per barrel of oil from this source is often quoted as being the same as the cost of oil from coal. Concomitantly, it might be appropriate (although not completely correct) to call the market price of conventional oil at which the actual building of the backstop technology is begun the *backstop price.*

The question now is: Just what kind of present oil price is implied in the light of the costs required to put an alternative oil-producing technology in place? Several years ago we were talking about an oil price of $14/bbl to $20/bbl as being sufficient to induce the large-scale production of oil from coal, by employing a technology developed in Germany during World War II. This figure is currently put between $35/bbl and $50/bbl, and lead times of up to fifteen years may be involved. In the United States in 1976-1977, some typical lead times for energy resources were as follows: new onshore and natural-gas fields, one to four years; existing offshore fields, principally in the Gulf of Mexico, three to seven years; fields in remote areas such as Alaska and the Atlantic and Pacific Oceans, six to thirteen years; plants for producing synthetic oil, five to ten years; nuclear and hydroelectric power plants, nine to thirteen years; other power plants, seven to nine years; new coal mines, four to ten years. These figures are probably higher now because of the increasing costs of environmental protection and the greater emphasis on safety that will almost certainly follow the incident at Three Mile Island (Harrisburg, Pennsylvania). For various reasons, including scarcity of data, the matter of lead times is not considered in what follows, but the reader should not forget that even if a decision were made today to expand synthetic-oil facilities at a maximum rate, it would be many years before they could contribute a sizable fraction to the world energy supply. The figure of five to ten years provided above is applicable only in the event that sufficient inputs of coal and so on are available; and securing enough of an item such as coal to support a major synthetic-oil program would, in itself, involve huge investments in not only coal mines, but also transportation facilities.

Given the present transition in OPEC's concern from prices to quantities (which is equivalent to a decrease in concern for present revenues and an increased interest in conservation), as well as the near certainty that on the basis of present consumption trends and information about reserves, the world production of petroleum will begin to decline before the year 2000, it seems reasonable that the present price of petroleum not only should be sufficient to cover the cost of the oil being produced today, but also must contribute to paying for the equipment that will produce a substitute for oil when this substitute is required—assuming that there is still a demand for an oil-like commodity. Note that this is the same terminology we use when discussing a physical asset: as a physical asset (for example, a machine) depreciates, enough money must be set aside that a new asset can be purchased. In what follows we talk in terms of replacing the commodity itself rather than the asset for producing it. For instance, in the calculation given below, the assumption is that in thirty years' time as much synthetic oil will be required as conventional oil being used today (which is probably an exaggeration). Thus each barrel of conventional oil sold today must yield not only enough to cover its own cost, but a "surplus" which, if invested in a bond or bank account, will grow at such a rate that in thirty years it can purchase a barrel of synthetic oil at the price at which that oil will be selling at that time, which is the present price adjusted upward at the rate of inflation.

This exposition will be sharpened slightly before we go to our calculation. If we have a sum of money, say \$100, and the rate of interest r_m is 10 percent, then in a year this money will have grown to $100(1 + r) = 100(1 + 0.10) = \$110$. But we can also take this in reverse! One hundred and ten dollars one year from now is \$100 today. What we say is that the *present value* of \$110 a year from now is \$100 dollars; and it should also be clear that this \$100 is the amount we must save in order to have \$110 one year from now. Similarly, with the same rate of interest \$100 now will grow to $100(1 + r)(1 + r) = 100(1 + r)^2 = 100(1 + 0.1)^2 = \$121$ two years from now. If we move in the other direction, two years from now \$121 has a present value of $121/(1 + r)^2 = 121/(1 + 0.1)^2 = \$100$ at present. Thus, if we have a sum of money x, in t years its present value is $PV = x/(1 + r)^t$. For instance, the present value of \$1,000 in 8 years is, with $r = 10$ percent, $\$1,000/(1 + 0.10)^8 = 1,000/2.1436 = \$466.50$. Put another way, this amount of money would have to be saved today in order to have \$1,000 eight years from now.

One more concept that must be understood concerns the inflation rate, or rate of price increase for goods and services, which is usually measured on an annual basis. If, for instance, we know that the general rate of inflation for industrial goods is 7.5 percent, then an estimate of how much a piece of equipment costing \$500 dollars today would cost in one year would be $500(1 + r_i) = 500(1 + 0.075) = 537.5$, where r_i is the inflation rate. If we were interested in the price of this equipment in ten years and if we thought that the inflation rate would remain unchanged or would average 7.5 percent or thereabouts, we would have $500(1 + 0.075)^{10} = 1,030.5$ as the estimated price of the equipment in ten years.

We can now try to get some idea of just how much a barrel of oil should cost today, assuming that in 30 years every barrel of conventional oil now being produced will have to be replaced by a barrel of synthetic oil. If we make the assumptions that a barrel of synthetic oil today costs \$40, the general rate of inflation in the industrial world is 7.5 percent, and the market rate of interest is 10 percent, then the present money price of conventional oil in a perfect market would be the present marginal cost of a barrel of oil plus

$$\frac{40(1 + r_i)^{30}}{(1 + r_m)^{30}} = \frac{40(1 + 0.075)^{30}}{(1 + 0.10)^{30}} = \$20$$

Of course, this is an approximation, with the principal shortcoming being, from my point of view, a lack of knowledge as to the exact amount of synthetic oil required in thirty years. However, there are two ways of attacking this problem. The first is to assume that the time span for replacing all oil by synthetic oil is fifty years instead of thirty. The above calculation then gives \$12.5 instead of \$20. This is much less, of course, but it is still much more than the marginal cost of OPEC oil (and is almost equal to the OPEC average price of 12.90 on 31 December 1978.

Then, too, instead of dealing in oil, we could concern ourselves with the energy content of oil. We would be replacing conventional oil by synthetic oil and, in some uses, other energy sources. As long as the same amount of energy were being replaced, and the other energy sources were at least as expensive as synthetic oil, the above calculation would hold. Other objections that could be brought against the above include the real yield on capital implied by the assumed market rate of interest (r_m) and rate of inflation (r_i). This is approximately $r_m - r_i$ and amounts to only 2.5 percent, which is unusually low, If, instead, we used 4 percent, then the above calculation would yield a price of \$12.3 [$= 40/(1 + 0.04)^{30}$]. *But,* on the other hand, it seems very dubious that synthetic oil could be produced on other than a pilot scale for \$40 dollars; \$50 would probably be a more appropriate figure to use in the above calculation, and the present rate of inflation is not 7.5 percent, but closer to 12.5, and so on. Rather than continue in this vein, it might suffice to say that until recently the market price of oil reflected, to a considerable extent, its actual scarcity; and given the uses to which this commodity might be put by the oil-producing countries in the future, it is not inconceivable that its present price is appropriate; and some analyses have begun suggesting a per barrel oil price in the vicinity of \$90 per barrel by 1990.

The Oil Price, Reserves, and Market Forms

The problem of the optimal rate of exploiting an exhaustible resource has been mentioned earlier in this book, but to make the ensuing discussion more comprehensible, those remarks need to be extended slightly. As before, the issue is simply whether a unit of a resource should be extracted today or remain in the ground where it can produce a current return for its owner by appreciating in value. The idea is that periodically the resource owner must ask the question put by Solow (1974): What have you done for me lately? This question is actually all-inclusive, but for pedagogical purposes it is reformulated: What have you done for me lately in comparison to what other assets are doing, given my appetite for current pleasures, and what are you going to do for me in the future?

To take the first part, resource deposits in the ground should be growing in value at a rate equal to the rate of interest if they are to be left in the ground. Otherwise they should be extracted and sold, and the revenue from these sales should be invested in a bond or savings deposit since, by definition, these particular financial assets *are* yielding the current rate of interest and, if left undisturbed, *are* growing in value at a rate equal to the rate of interest. What about current pleasures—wine, watermelon, and song? Since in a textbook situation the market rate of interest and the subjective discount rate are equal, anyone believing that current consumption should be postponed only if, by doing so, future consumption could be increased, would require the asset to increase in value in the ground by an amount equal to the discount rate (which is the premium required for relinquishing current pleasures in favor of future gratification). If, for example, the resource owner has a subjective discount rate of 10 percent (which in real terms means giving up 10 units of consumption today only if the owner would thereby gain 11 units in a year's time) and if the resource is appreciating in the ground at a rate of only 5 percent, then the resource is *not* doing enough for its owner in its present state. Therefore the owner would undertake to transform at least some of it into bank accounts, town houses, ski trips, and steak dinners. The reader should also realize here that "appreciating in the ground" refers to the gain in value of the resource due to an increase in the actual or anticipated price of the resource.

The basis of this analysis is the well-known condition for equilibrium in asset markets: there can be equilibrium only when all assets in a given risk class earn the same rate of return, partly as current dividend and partly as capital gain. (In appendix 4A this definition is used to provide a new mathematical statement of the discussion in the previous paragraph, the so-called Hotelling condition.) Among other things, this means that oil in the ground often can be considered a more valuable asset for OPEC than other assets which could be bought for oil revenues. The two most important reasons are as follows: (1) given the present rate of price inflation in the industrial countries, most financial assets yield a real return of only 1 or 2 percent, if that, and many physical assets purchased by OPEC, such as plant and machinery, often do only slightly better; and (2) if the OPEC countries cooperate and do not lift too much oil, not only will the price of that which they produce be high, but also a gratifying rate of price increase can also be ensured, thanks to the structure of the oil market and the procedures for setting the price.

There is also the matter of uncertainty, or, what can you do for me in the future? In Keynesian macroeconomics, money kept inside the mattrress and in cookies jars, which earns no return at all, is not removed and transformed into stocks, bonds, and real estate because of uncertainty about being able to convert these investments to cash later in the event of an emergency or chance to participate in some advantageous transaction or speculation in the future. The same thing is true here. When oil in the ground is gone, it is gone forever, and so are the power and prestige of its owners unless they have replaced it by suitable industrial and educational assets, which is not always easy to do. But no country or government which can look forward to possessing an exportable surplus of "black gold" need be overly concerned about its ability to generate a satisfactory reward during the next few decades, because mistakes made when calculating the benefits from present investments can be compensated in future price-fixing sessions or, as is now becoming the fashion, through individual sellers placing ad hoc surcharges and levies on their product which buyers inevitably pay in order to help guarantee future supplies. Apropos the term *black gold,* often applied to oil in those days when it was singularly inexpensive, it is ironic to note that oil is in the process of becoming a new international standard of value, replacing both the auric metal and dollars.

Now we can turn to the work of Pindyck (1978), which probably contains the best mathematical model available for the examination of intertemporal resource allocation in a cartelized natural-resource industry. Pindyck's model takes specific account of the exhaustion of the resource, as well as the oligopolistic structure of the world oil market, specifying both a cartel and a "competitive fringe." The framework of the analysis is the dominant-firm model as clarified by Folie and Ulph (1978, 1979, 1980). Here the cartel sets a price for the resource, and the fringe accepts the price and selects its output. The cartel then meets the residual demand, which is equal to total demand minus fringe supply. At the same time, though, it should be noted that within the cartel, Saudi Arabia seems to be playing the part of the dominant firm, adjusting its production to meet that part of the residual demand not covered by the other members of the cartel. This is an important matter, and it is taken up in some detail later.

Right now, however, we want to stress that aspect of Pindyck's work which emphasizes that the short-run elasticity for oil is considerably lower than the long-term. For instance, if the price of oil rises by a certain percentage, we expect the demand to fall by a certain percentage, but in the short run the decrease in demand will be very small, since oil is virtually an indispensible industrial input. In the long run, however, a certain amount of oil can be dispensed with either through savings or through changing the design of the equipment and structures using oil. It also appears that a stable relationship exists between energy and income: A 1 percent increase in income brings about an increase in energy demanded of 0.80 to 1.1 percent, depending on the country. As alluded to above, Pindyck's model also postulates a profit-maximizing behavior which explicitly takes into consideration the effect of using up reserves today on future profits: Even if the revenue from oil production exceeds the sum of fixed and variable costs by a considerable amount during any given period, an increase in production could still result in a fall in the present value of discounted profits because of the using up of reserves that are the source of future earnings.

All these things together suggest a strategy that turns on raising the price of oil sharply from time to time and then holding it constant. The very low elasticity of demand guarantees higher revenues, and although these may tend to decrease over time for the reasons given, this negative effect is largely mitigated by the increase in demand (or shifting of the demand curve to the right) resulting from the growth of incomes in the oil-consuming countries. Basically the idea is high revenues *and* a low rate of depletion of reserves. The only finesse that the cartel needs to display involves making sure that the rise in oil prices does not completely inhibit the economic growth of their clients or lead to unpleasant economic or political countermeasures. It has also become evident that these sudden price escalations arrest the falling real price of oil. The real price of oil is explained in chapter 2 and is touched on again in chapter 6 in conjunction with an extended discussion of oil and international monetary economics. For the purposes of this discussion, the *real price* of oil can be defined as the money price of oil divided by the general rate of inflation in the major oil-consuming countries, which are also the countries from which OPEC purchases most of its industrial goods.

The most interesting examinations of Pindyck's model are undoubtedly those of Folie and Ulph (1978, 1980). Among other things, they have shown the economic viability of a policy which calls for fixing the output of oil and letting the price rise to clear the market. Employing Pindyck's model with a time horizon stretching to the year 2010 and a fixed production of 7.5 Gbbl/yr, they demonstrate that discounted profits fall by only 6 percent in respect to the profits that would have been obtained in Pindyck's version of the optimal policy (which considers only profits), but at the same time OPEC's reserves in 2010 are doubled. The thing to understand here is that given the factor of uncertainty, particularly in regard to the future demand for oil by the OPEC countries themselves, some weight should be given to reserves in their own right. Depending on how this is done, perhaps a program which reduces "estimated" profits by only 6 percent, but results in a very large amount of reserves still being available in 2010, is, in fact, superior to the optimal program in Pindyck's model where only discounted profits are considered.

We can now say something about Saudi Arabia, which apparently plays the part of the dominant firm, or residual producer, where OPEC is concerned. In a sense this makes Saudi Arabia the residual producer for the entire world, covering a major part of the difference between the total world demand and production in the other OPEC countries, North America, the centrally planned countries, and so on. The problem, it has been maintained, is that playing this role can involve having to make some painful production cutbacks from time to time: in 1978, when so much oil was reaching Europe that it could not be stored, not only Saudi Arabia but also Kuwait and the United Arab Emirates were producing at less than 50 percent of total capacity. The argument is presented now that these and other countries currently may be in a situation in which they will find it more unsatisfactory to permit annual increases in their output than to allow a substantial portion of their production facilities to remain idle.

First, consider a simplistic analysis based on the work of Lichtblau (1977). Only a short time ago some of the most respected and experienced oil analysts were insisting that Saudi Arabian oil production was on its way to a maximum output of 18 to 20 million barrrels per day (Mbbl/d) by 1990. The reader can find a more thorough comment on the circumstances surrounding this prediction in my book *The International Economy: A Modern Approach* in which the reaction of the Saudi Arabian oil minister, Skeikh Yamani, is registered. But it is perfectly clear now that this was unvarnished daydreaming, brought on by the knowledge that, *ceteris paribus,* a failure by Saudi Arabia to adjust its production so as to move toward the above forecast could mean a palpable fall in the aggregate growth rate of the industrial countries. As explained in chapter 2, this, in turn, would mean an exacerbation of the already high rate of unemployment in many of these countries, less investment, more inflation, less international trade, and almost certainly a decline in living standards for tens of millions of people in Europe and North America. Since it now seems likely that by 1990 Saudi Arabia will be producing only slightly more than one-half of the desired levels, it is probably a good idea for many people in the industrial countries to start setting their consumption sights a little lower, if they have not already done so.

The basic dilemma is that Saudi Arabia cannot possibly cooperate in making the dreams of its customers come true if it is cognizant of its own long-run economic interests, which apparently it is. If Saudi Arabia were to increase its production to 18 to 20 Mbbl/d by 1990, its reserve/production ratio would drop to about 17, assuming that no new oil were discovered in that country. But even if as much oil were discovered as is extracted in order to meet the target of 18 to 20 Mbbl/d Saudi Arabia's reserve/production ratio still would be down to almost 25 in the same year. (To see this, we have only to understand that if as much oil is discovered as is extracted, reserves are unchanged. But production is growing, and thus the *R/P* ratio falls.) Then, if reserves were to fail to expand in pace with production, which would be almost certain under the circumstances, an output of 18 to 20 Mbbl/d could lead to a very rapid decline in the *R/P* ratio. In order to arrest this tendency, production might have to be cut back by a very large amount, which could result in the idling of a large amount of very expensive equipment that was only recently purchased (in order to get production up the the level of 18 to 20 Mbbl/d).

The reader should also attempt to appreciate the fact that regardless of the huge investments made in Saudi Arabia over the past few years, the country is still decades from even the first stage of a takeoff into self-sustaining growth; and if it is the sincere intention of the Saudi leadership to trade oil for development, they will simply be in no position to tolerate an accelerated fall in their *R/P* ratio. Although an *R/P* ratio of 20 to 25 may seem more than adequate for a country such as the United States (where this ratio is now under 10), it may not suffice for a less developed country with serious development ambitions, but no industrial base or access to a large supply of trained workforce.

As far as the undiscovered reserve potential of Saudi Arbia is concerned, there are plenty of theories which insist that Saudi Arabia's undiscovered reserves are huge. However, some of the companies in Aramco (the Arabian-American Oil Company, of which 60 percent is owned by the Saudi Arabian government and 40 percent by a consortium of Exxon, Texaco, Standard Oil of California, and the Mobil Corporation) claim that there are only modest amounts of undiscovered reserves in Saudi Arabia. One company, for example, believes that there is an undiscovered reserve potential of only 33 Gbbl of oil in that country. The key question for this book, though, is: How much will be discovered in the coming ten or fifteen years? And, as pointed out in chapter 3, regardless of how much there actually is, the government of Saudi Arabia is in no particular hurry to locate it.

Equally important, there is the possibility of large pressure decreases being brought about by a production growth of the magnitude given above. Here it should be mentioned that a recent report of Aramco stated that a Saudi production of 16 Mbbl/d could be maintained for only a maximum of ten years before such things as pressure deficiencies would impose a sharp decline in output. Putting all these things together indicates that Saudi Arabia may reach and maintain the capacity to produce as much as 11 to 12 Mbbl/d of oil, but average daily production generally will be less. In fact, it is said that there are influential persons in that country advocating a production of as low as 7 to 8 Mbbl/d and that there is a growing receptivity to this position. After all, this invaluable wasting asset is not being replaced by industrial and educational assets of equivalent value, and it is apparently an open secret in the Middle East that projects undertaken in most of the oil-producing countries in that part of the world are two or three times as expensive as comparable projects in developed countries.

We now come to an interesting extension of this analysis which involves differential field size and the fact that in Saudi Arabia if production is to be raised by a generous amount above the current level of 8.5 to 9.5 Mbbl/d, much of this increment would have to be taken from the world's largest oil field, Ghawar, which has 10 percent of the world's remaining reserves and currently accounts for more than 50 percent of Saudi Arabian capacity and 37 percent of its reserves. But because of the economic-rate-of-recovery constraint mentioned in chapter 3, as well as various pressure problems, the Ghawar field can support a large increase in output for only a fairly short time. After only a few years production would begin to fall, with the subsequent loss of a large part of the very expensive investments required to build up capacity.

To get some idea of the issues involved here, we postulate a model economy with three oil fields. Two of them are small, with reserves of 60 units each at the time we begin our exercise. The third field is large, with reserves of 200 units. Assume also that production in each of the small fields has been one unit per year for a long time in the past, whereas for the large field it was 5 units per year. At the moment of beginning our exercise, the decision has been made to raise total production by 7 units, and the question is immediately raised as to how this can be accomplished. For one thing, there seems to be an infinite number of strategies. One which comes immediately to mind calls for raising production in each of the smaller fields by 2 units per period (from 1 to 3 units) and in the large field by 3 units per period (from 5 to 8 units). But as is made clear in figure 4-1, this strategy causes problems. If we assume a critical R/P ratio of 15, then production will have to decline immediately in the smaller fields (because the critical R/P ratio is reached). Thus we waste a large part of the investment needed to reach that capacity. Moreover, as the reader can easily verify, the ensuing rapid rise in production in the larger field means that we will soon run into the critical ratio there also, and production will have to decrease very sharply.

Another arrangement, which is analogous to the one Aramco thinks most economical for Saudi Arabia in the event that it decides to produce 15 to 16 Mbbl/d, would be modest increases in the smaller fields (say from 1 unit per day to 2 units per day) and a substantial increase in the large field (from 5 to 10 units per day). This scheme is shown in figure 4-2. Once again, when the R/P ratio reaches its critical value, which in this example is 15, this critical value "takes over" and determines production. Before we look at the arithmetic of this example, note that many arrangements superior to the one proposed here are conceivable, all that I have examined require large increases in the output of the large field relative to the size of increases in the smaller field and give basically the same result obtained in this illustration.

Now we can make some calculations of the R/P ratios. These are also shown in figure 4-2. In year 6 we reach the critical R/P ratio, (15), for the large field and then that ratio takes over, as described. For instance, in the following period, with reserves at 140 units, production is determined by the critical R/P ratio and is equal to $140/15 = 9.33$. At this point the other two fields each have R/P ratios of 24. The aggregate R/P ratio for the country when production turns down is thus

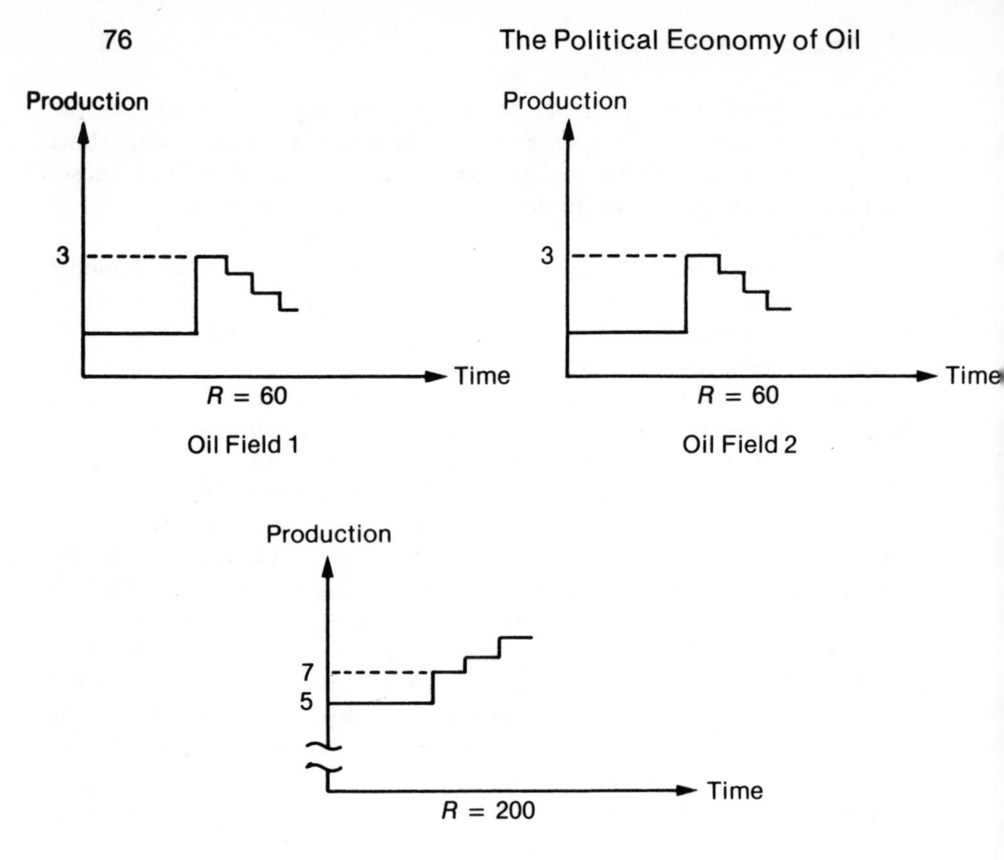

Figure 4-1. Some Theoretical Production Patterns in a Model Economy with Three Oil Fields, First Postulate

$$\frac{R}{P} = \frac{48 + 48 + 140}{9.33 + 2 + 2} = \frac{236}{13.33} = 17.70$$

The thing to discern here is that the aggregate R/P ratio is greater than the critical ratio (which is 15), but even so production has turned down. Of course, production could be kept up by gradually raising output in the smaller fields, but the issue then becomes whether it would be profitable to do so in light of the investments that would be required. Although we will not do so in this book simple examples can be constructed to show that with the figures used above, and cost curves of the usual type, this is not the case.

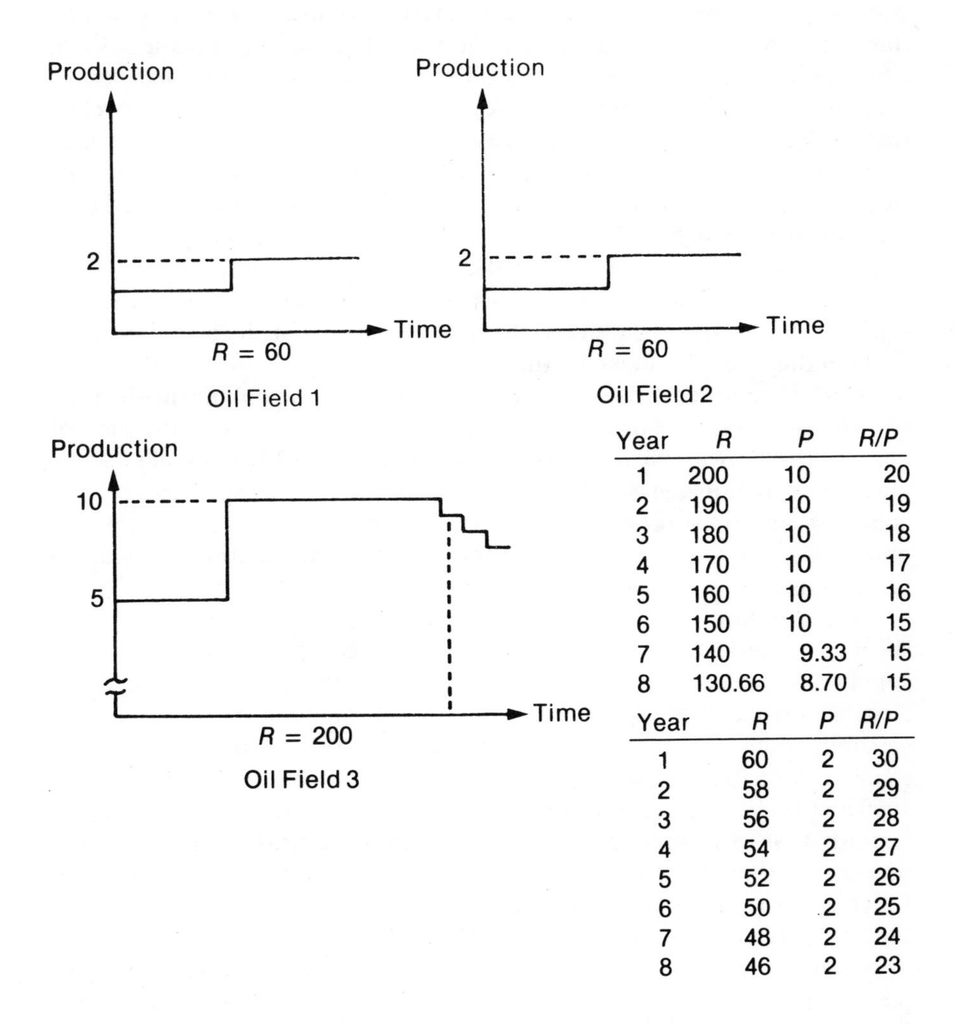

Year	*R*	*P*	*R/P*
1	200	10	20
2	190	10	19
3	180	10	18
4	170	10	17
5	160	10	16
6	150	10	15
7	140	9.33	15
8	130.66	8.70	15

Year	*R*	*P*	*R/P*
1	60	2	30
2	58	2	29
3	56	2	28
4	54	2	27
5	52	2	26
6	50	2	25
7	48	2	24
8	46	2	23

Figure 4-2. Some Theoretical Production Patterns in a Model Economy with Three Oil Fields

Price and Inventories

We turn now to the link between prices and inventories, but first we must consider a few preliminary matters. The term *spot price* has been used, and this is also occasionally called the cash price. The *cash price* is the price at which goods are for sale at a given time. There is also the *spot market;* an example is the Rotterdam spot market. It should be emphasized, however, that the spot market is not necessarily a market, but an arrangement which calls for a transfer of ownership from seller to buyer in the immediate future. (The arrangements for delivery may also be carefully specified on a spot contract.) The *forward market,* however, involves *forward* sales, or the sale of an item to be delivered in the future at some mutually agreed-on price (for example, the official or *posted* price of OPEC at the time the deal is made). But it could conceivably happen (and undoubtedly does from time to time) that the price at which the commodity changes hands is the price at or around the time of delivery, and this price may be unknown when the transaction is made. For instance, the price may be higher at the time of delivery than it was when the commodity was purchased. It should also be made clear that, in general, a forward market involves physical delivery and is not the same as a *futures market,* in which delivery takes place in only a minority of cases and, as is made clear at the end of this section, the majority of transactions are paper transactions.

We should also make some effort to distinguish between long- and short-run prices. Long-run prices are determined by trend movements in supply and demand. If over a long period the capacity to supply an industrial raw material expands at a more rapid rate than the demand, then generally there will be an ineluctable downward pressure on price. This was a major contributing factor to the constant oil price experienced between 1960 and 1969, despite the very rapid growth in the world economy and particularly that of Europe. New producers who, in contrast to the major oil-producing firms, were known as independents began to exploit reserves in Libya and elsewhere, and there was a record acceleration in production in the Soviet Union and Saudi Arabia. Here the reader should be careful to note that the constant price of which we speak is a *money* price. During this period the real price of oil, of nonfuel minerals, and of agricultural products fell, which is a very important factor in explaining the prosperity achieved by a large part of the industrial world *and* by some of the newly industrializing countries in the Third World.

But short-run prices are characterized by their volatility and under close examination often reveal peaks and troughs that are separated by weeks or months, instead of years; and they are, to some extent, independent of business activity. A large part of the explanation for these particular price oscillations are speculative tides of bullishness and bearishness fueled by fantasy, naiveté, or bad judgment. Thus a relatively minor surge in supply might make oil prices fall for a few weeks, causing an influential group of market analysts to glimpse what they think is a surplus situation. Something like this happened in 1978 when prices on the (Rotterdam) spot market fell below OPEC's posted price. The failure of this price to conform to the expectations of many analysts and remain in that position was, however, no surprise to those familiar with the market power of OPEC vis-à-vis the rest of the world, although those opinions were pretty well ignored by the cognoscenti.

It is not very far from considering the variation in short-term prices to the subject of inventories (or stocks). Among the motives for holding stocks is speculation, where the ability to perceive the turning point in price cycles and to formulate valid price expectations is an invaluable aptitude. Also in many markets rapid changes in the level of stocks held for speculation can destabilize a market by increasing the amplitude of price swings. Concomitantly, much of the information contained in prices can be lost, in the sense that the "true" market situation is obscured.

Speculation can occur for several reasons and in a variety of ways. A *long* position can be taken in a commodity, which means that it is bought, for example, on the spot market, and held for the purpose of a later sale in expectation of making a capital gain. Another way of speculating is to be *short* in a commodity, which works as follows. A speculator agrees to sell a commodity in the future for a price just above the price at which she or he "thinks" the commodity can be bought just before the day on which the purchaser is to receive the commodity. In other words, when the (forward) contract comes due, the speculator buys on the spot market for a price that, if all goes as planned, is under the sales price on the forward contract. The term *short* simply means selling something that one does not have.

Stocks can also be held for insurance purposes. If a producer's or a consumers' inventories are low, then each extra unit held in stock reduces the possibility that production will have to be scaled down because of an unforeseen absence of some input. Remember that both producers and purchasers of industrial raw materials are bound by contractual obligations to their customers; as a result, inventories must be held even if there is an inverted relationship between the spot price of the commodities being stocked and all predictions of the future price. Even if the expected money yield from holding and later selling a commodity does not cover such things as its storage cost, this negative aspect is balanced by a positive *convenience yield* when the size of inventories is small relative to the amount of the commodity being used as a current input in the production process. In this situation an effective price system must function to ration existing stocks among existing and potential stockholders, and often this calls for a departure from normality in the form of an inversion between present and future prices.

This inversion is called *backwardation,* and it might be possible to speak of it existing on the world oil market at the end of 1979 and beginning of 1980 as a result of some misunderstandings about the future availability of oil. At the Caracas meeting of OPEC in December 1979, Sheikh Zaki Yamani pointed out that the spot price of oil would almost certainly fall in 1980, and there were few who disagreed with this prediction. Although world inventories at that time were almost at a normal level, they were not considered large enough, given the uncertainty prevailing on the oil scene after the change in government in Iran, and so a large amount of very expensive oil was purchased on the spot market to add to these inventories.

By the same type of reasoning, if the amount of stocks being held is large in relation to the amount being used as a current input in the production process, then there is little incentive to held more. In these circumstances, convenience yields are small, and stockholders require that the expected future price of the commodities being held cover storage handling, insurance, and other charges—unless it is possible to sell the commodity forward at a premium which, in the opinion of the stockholder, would be sufficient to cover carrying costs. Otherwise these stocks are put on the market, driving down spot prices and widening the gap between present and expected future prices to the extent that holding existing stocks is justified. Something like this may be happening on the world oil market at present, and will very likely continue for a while, as the "extra" inventories purchased as protection against an oil shortage of the kind that threatened when exports from Iran were reduced by up to 2 Mbbl/d in 1979 are sold. Figure 4-3 illustrates some of these concepts for an unspecified commodity.

In the diagram p_s is the spot price, and p_f is the *expected* future price of the commodity. A proxy for p_f might be the actual price of forward contracts on an organized commodity exchange (such as the London Metal Exchange). At *Z* we see the shift from backwardation at low levels of the inventory/consumption (*I/C*) ratio for this commodity to the normal condition called *contango,* where the expected future price exceeds the spot price. To relate this figure to a real-world situation, note that a year or so ago copper inventories amounted to four to six months of the annual consumption of that metal. Thus when fighting broke out in Shaba (Katanga) province in Zaire, the fall in world inventories corresponded to a movement from *B* to *A* in the figure. With expectations about future prices unchanged, this fall in the *I/C* ratio was hardly enough to cause a ripple in the quoted prices of either the metal or the ore.

In closing this section, we discuss another type of price and market, since it has been suggested that some of the problems endemic to the oil market could be solved if conditional futures markets were available on which risk could be traded. Futures markets exist for many commodities, although they do not operate so as to cover all contingencies. Houthakker

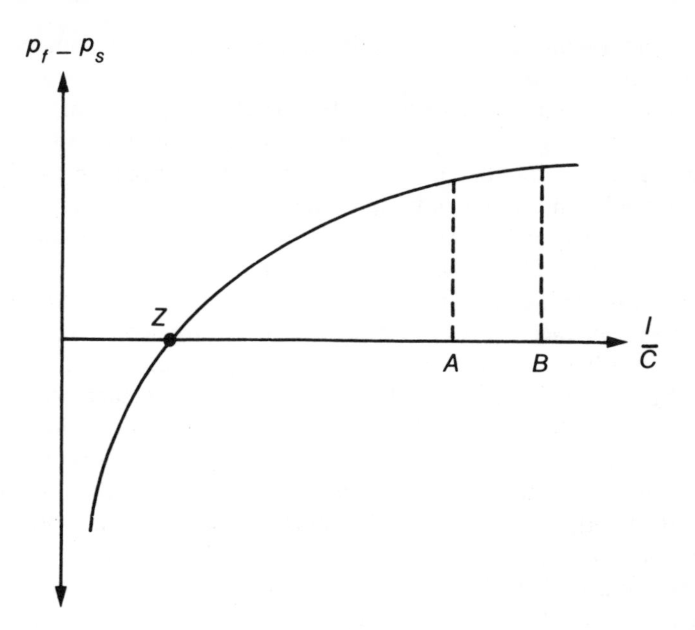

Figure 4-3. The Spread between the Expected Future Price and the Spot Price as a Function of the Ratio of Inventories to Consumption

(1973) has indicated, and incorrectly, that security and asset markets can ensure that prices in the oil market, for example, will not diverge markedly from the short-term efficient price; and they will also assist in reducing the instability of the spot market. By the same token Houthakker seems to be saying that the absence of futures markets does not represent too great a loss. The feeling here, though, is that any arrrangement which transfers risk from those who do not want it to those who do, represents an unambiguous welfare gain. At the same time a futures market for oil may be out of the question for technical, as opposed to conceptual, reasons. This matter is taken up below.

As for exactly what a futures market is, the first point to note is that it is concerned with *hedging,* or transferring price risk from buyers and sellers of physical commodities to *speculators.* Next, a futures market features delivery in only a minority of cases. Strictly speaking, most (if not all) futures contracts *are* forward contracts and stipulate the delivery of physical commodities at some point in the future, but this market is organized so that sales or purchases of these contracts can be offset and future deliveries are unnecessary. The key element in this scheme is the presence of speculators who buy or sell futures contracts with the intention of making a profit on the difference between sales and purchase price. The details of this procedure can be found in Banks (1979a). In brief, the arrangement might function as follows for the case of someone buying a physical commodity.

Knowing that he or she has to buy a certain amount of the commodity in a month or two, this person would hedge the price risk by buying a futures contract for the commodity. Then at the point in the future when the physical commodity is bought, the futures contract is sold. If we are thinking in terms of oil and if we can assume that a futures market in oil (or oil products) would function in the same manner as one for copper or agricultural products, then an increase in the price of the commodity would also mean an increase in the price of the futures contract. Thus the loss taken if we have an increase in the price of the commodity would be canceled by being able to sell the futures contract at a higher price than that at which it was bought. Notice also that if the price of the commodity fell, so should the price of the contract, and thus the good news on the commodity market would be canceled by the bad news on the "paper" or futures transaction. Another interesting variation on the same theme would be a dealer buying a commodity which would be kept in inventory for several months before being sold. If this dealer did not want to speculate i.e., gamble on a price rise, he could hedge his price by selling a futures contract. Then, at the same time as he sold the commodity, an "offsetting" purchase could be made in the futures market. *If* the commodity did fall in price, the same thing should happen to the futures contract, and thus the loss on the physical transaction would be counterbalanced by a gain defined as the difference between the selling and buying prices of the futures contract.

What has taken place in these examples is that dealers in physical commodities have turned over a large part of their price risk to speculators, who buy and sell futures contracts with the intention of making a profit on the difference between buying and selling price, or vice versa, depending on whether they think the price of the commodity will rise or fall. Basically, the dealer is concerned with insurance and regards the buying and selling of futures contracts as being of secondary concern: in the long run, with extensive buying or selling of these contracts, any small losses that he might take on individual contracts would be canceled by small gains. Speculators, on the other hand, are gamblers: some make money, while others eventually retire to make room for other losers.

Just recently the London Commodity Exchange began promoting the idea of an international petroleum futures market which, to their way of thinking, would help reduce the volatility of the Rotterdam spot market. (Of course, it does not necessarily follow that this would happen.) One of the problems that has not been adequately dealt with in this proposal concerns the matter of physical delivery, which is a feature of all existing futures markets. The *London Economist* (April 26, 1980) has briefly reviewed the merits and demerits of such a scheme, and concluded that without physical backing, futures contracts in oil would be unsatisfactory. In this particular matter, I do not believe that the problems that would be encountered with oil would be more complex, in principle, than those found in the futures market for gold. Large inventories of oil certainly exist in private hands; and even on some organized futures exchanges, delivery does not take place directly to clients, but to a limited number of warehouses.

However, a mammoth problem would have to be faced if the most important oil producers decided to be active in such a futures market, since by withholding or dumping physical supplies at the appropriate time they could ensure that all their "bets" on the futures market paid off. Suggestions have also been forwarded for an exchange that would not trade in oil itself (because of the difficulty of establishing a standardized contract), but in gas, oil, and naphtha. In my opinion there are a number of possibilities here, but a great deal of thought will have to be given the design of the contract.

Concluding Remarks

Finally, it should be pointed out that there is no shortage of people who believe that the present world oil crisis could be radically reversed in a short time. Under certain circumstances, I would be willing to accept 50 percent of this position: eventually, either through the discovery of supergiant fields in relatively unexplored or hard-to-reach regions or through a rapid increase in the availability of alternative energy resources, the buyers of oil might gain the upper hand over the sellers. But even if this were to happen, it would not happen soon, and the short rather than the long run is important where this dilemma is concerned.

There is also a determined body of opinion which takes every possible opportunity to assert that Mexico will soon be able and willing to compensate any shortfalls in the output of oil that may result from various OPEC members showing a less pronounced willingness to collaborate with the industrial countries in the squandering of this invaluable resource on trifles, and nonsense. Occasionally we hear that Mexico has potential reserves of 150 to 300 Gbbl of oil (which would put it in a class with, or above, Saudi Arabia) and that by 1987 conceivably Mexico will be producing at least 7.5 Mbbl/d. Whether all this is true remains to be seen. According to Stobaugh and Yergin (1979), drawing on the extensive investigations of the Harvard Energy Project, proved Mexican reserves are only a fraction of the above figures. Moreover, according to information circulated recently by the Mexican government, former U.S. Department of Energy Secretary Schlesinger, and the International Energy Agency, Mexican oil exports probably will come to less than 3 Mbbl/d by 1985, regardless of their supply of recoverable oil.

But even if we assume that Mexico desires to and could produce 7.5 Mbbl/d in the late 1980s, if OPEC accentuates its present trend toward conservation, it would still control enough of the market to preserve almost all its present revenues and influence on the world oil price. Furthermore, remember that Mexico is a country with serious development ambitions, with no particular gratitude toward the industrial countries (and particularly the United States), and with an extremely high population growth. Between 900,000 and 1 million young people enter a job market every year that has no potential whatever to absorb more than half of them. Mexico's need for additional export revenue is going to be very great for a long time; and until some of the hypothetical reserve figures given above are substantiated, maybe the best way to obtain this revenue is to actively assist in keeping the price of oil as high as possible. Finally, despite noble insistances that oil is capable of turning Mexico into a modern industrial state, what it has done thus far is increase the gap between the haves and have-nots and expand the ever-present corruption in that type of society to elephantine proportions. All things considered, it might be best to not look forward to any tendencies toward price moderation from Mexico or, for that matter, any other oil-producing country anywhere.

Appendix 4A: Exhaustible Resources (Hotelling's Rule) and Price Formation in a Stock-Flow Model

An asset, K, is valued at time t according to the discounted value of its quasi rents, or

$$p_tK_t = \frac{p_{t+1}q_{t+1}}{1 + r} + \frac{p_{t+2}q_{t+2}}{(1 + r)^2} + \ldots + \frac{p_{t+i}q_{t+i}}{(1 + r)^i} + \ldots \quad (4A.1)$$

where p is net price (or price minus cost), q is production and r is the highest rate. But notice that the valuation of the asset at time $t + 1$ is

$$p_{t+1}K_{t+1} = \frac{p_{t+2}q_{t+2}}{1 + r} + \frac{p_{t+3}q_{t+3}}{(1 + r)^2} + \ldots \quad (4A.2)$$

Multiplying both sides of equation 4A.2 by $1/(1 + r)$ allows us to write equation 4A.1 as

$$p_tK_t = \frac{p_{t+1}q_{t+1}}{1 + r} + \frac{p_{t+1}K_{t+1}}{1 + r} = \frac{p_{t+1}q_{t+1} + p_{t+1}K_{t+1}}{1 + r}$$

And since $K_{t+1} = K_t - q_{t+1}$ for an exhaustible resource, we get

$$p_tK_t = \frac{p_{t+1}q_{t+1} + p_{t+1}(K_t - q_{t+1})}{1 + r} = \frac{p_{t+1}K_t}{1 + r} \quad (4A.3)$$

From this we immediately get the well-known relationship first derived by Hotelling, assuming that the asset is the stock of an exhaustible resource:

$$\frac{\Delta p}{p} \equiv \frac{p_{t+1} - p_t}{p_t} = r \quad (4A.4)$$

Observe the timing here. Production in the period from t to $t + 1$ results in an output called q_{t+1}, and this is sold at the price at which the stock is valued in period $t + 1$. But what happens if we sell the output at the price at which the stock is valued in the present period? This price seems reasonable since the price we would most likely use in discussing the value of the stock would be that associated with the most recent sale of output. (In a private communication Michael Hoel states that he believes it best to value whatever one keeps of the resource at the same price one obtains when selling the resource.) Now, calling production (which still takes place during the same period) q_t, we get

$$p_t K_t = \frac{p_t q_t}{1 + r} + \frac{p_{t+1} K_{t+1}}{1 + r} = \frac{p_t q_t}{1 + r} + \frac{p_{t+1}(K_t - q_t)}{1 + r} \tag{4A.5}$$

or

$$\frac{\Delta p}{p} \equiv \frac{p_{t+1} - p_t}{p_t} = \frac{r}{1 - \frac{q_t}{K_t}} \tag{4A.6}$$

Notice how the very important ratio q_t/K_t modifies Hotelling's rule. As is obvious from the above discussion, this is the production/reserve ratio, and as it rises ceteris paribus, the price of the resource should increase. It must be recognized, though, that the difference between equations 4A.4 and 4A.6 would vanish if we were dealing with continuous time. Similarly, we notice that if production is zero, we have $p_{t+1} = p_t(1 + r)$, and in equilibrium this is what we would expect.

Next we derive a simple relationship that gives some insight into the type of instability we normally see in short-run price relationships. For a detailed diagrammatic exposition of this matter, see Banks (1977, 1979b), but a sketch of the kind of model we are dealing with is shown in figure 4A-1.

The basic hypothesis here is that price changes are initiated by the demand for inventories not being equal to the supply, a position implicit in most econometric demonstrations. We can therefore begin with

$$p_t = p_{t-1} + \lambda(D_t - S_t) \tag{4A.7}$$

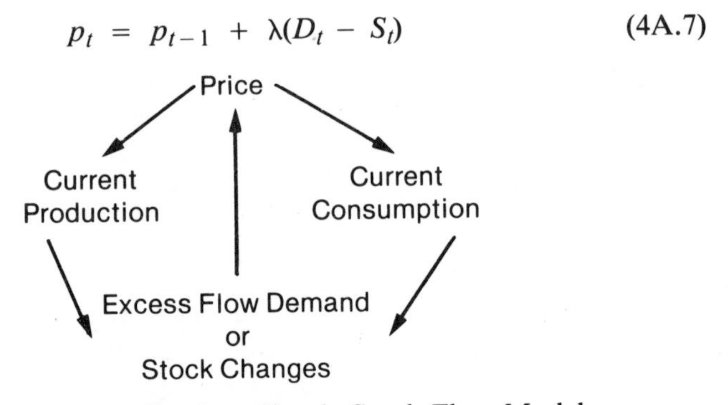

Figure 4A-1. Causality in a Simple Stock-Flow Model

where D and S indicate the demand for and supply of inventories, respectively, and $\lambda > 0$. We can now lag this equation to get

$$p_{t-1} = p_{t-2} + \lambda(D_{t-1} - S_{t-1}) \tag{4A.8}$$

Subtracting equation 4A.8 from equation 4A.7 yields

$$p_t - p_{t-1} = p_{t-1} - p_{t-2} + \lambda [D_t - S_t - (D_{t-1} - S_{t-1})]$$

or

$$p_t - 2p_{t-1} + p_{t-2} = \lambda [D_t - S_t - (D_{t-1} - S_{t-1})] \tag{4A.9}$$

Observe that $S_t = S_{t-1} + X_t$, or stocks in period t are equal to stocks in period $t - 1$ *plus* excess flow supply (where excess flow supply X_t is defined as *current* supply minus *current* demand). We can now substitute $S_t = S_{t-1} + X_t$ into equation 4A.-9 to get

$$p_t - 2p_{t-1} + p_{t-2} = \lambda (D_t - D_{t-1} + X_t) \tag{4A.10}$$

Next we need an expression for $X_t = s_t - d_t$. For simplicity, all supply and demand curves in this discussion are taken as linear, and so we have $s_t = e + fp_t$ and $d_t = g + hp_t$. In addition, let us take $D_t = \Theta + \alpha p_t$, which implies that $D_{t-1} = \Theta + \alpha p_{t-1}$. Thus equation 4A.10 becomes

$$p_t - 2p_{t-1} + p_{t-2} = \lambda [\alpha (p_t - p_{t-1}) + e - g + p_t(f - h)] \tag{4A.11}$$

or

$$p_t[1 - \lambda (f - h + \alpha)] + p_{t-1}(\lambda\alpha - 2) + p_{t-2} = \lambda (e - g) \tag{4A.12}$$

This is a simple second-order differential equation of the type

$$p_t + a_1p_{t-1} + a_2p_{t-2} = a$$

and it can be solved easily. Obviously the roots of this equation could be complex, which would mean oscillatory price movements. In addition, we could complicate these oscillatory movements considerably if we introduced some stochastic elements into these equations or slightly altered the lag structure, or both.

In conclusion, note that our equilibrium (when it exists) is the situation in which $p_t = p_{t-1} = \ldots = \overline{p}$. Thus we get from equation 4A.12

$$\overline{p} = \frac{e - g}{f - h} \tag{4A.13}$$

This is our condition for a flow equilibrium, with current supply equal to current demand. But we also see from equation 4A.7 that we have $0 = \lambda(D_t - S_t)$, and since $\lambda \neq 0$, we know $D_t = S_t$. This signifies that we also have a stock equilibrium, with demand for inventories equal to the supply.

Once again an oversimplified illustration of the backstop technology issue is in order. Take a case where we have two processes for obtaining oil. The first involves conventional oil, of which we use A units per year, and which can be extracted at zero cost. Assume that we have a total supply of B units of this oil. Let us also postulate an infinite potential supply of synthetic oil whose production requires nondepreciable capital worth K monetary units per barrel, but no other inputs. The rate of interest is r. (Notice that in this example the effect of price on future demand is sidestepped by specifying a yearly use rate of A). Now, with these data, the first units of synthetic oil (the backstop technology), will be required in $T = B/A$ years.

In a perfectly competitive market, at *switch point* T, the per period price of synthetic oil is equal to its cost, or $P = rK$. (This is the rental rate of nondepreciable capital; and no other costs for synthetic oil are assumed). Thus, with no extraction costs for conventional oil, the (efficiency) price of *conventional* oil in monetary units per barrel should be:

$$P(t) = P(T)e^{r(T-t)} = Ke^{-r(T-t)} \tag{4A.14}$$

This is the *scarcity royalty* on conventional oil, since with a positive demand for some kind of oil after conventional oil has been exhausted, sufficient cash must be accumulated to pay for the equipment to produce synthetic oil. By charging this price and putting the revenue from oil sales in the bank, or using it to buy bonds, by the time the existing oil stock is exhausted enough money will be available to purchase equipment that will indefinitely produce A units of oil per year. Note also that:

$$\frac{dP(t)}{dT} = -r\,Ke^{-r(T-t)}$$

If T increases (A decreases or B increases), then P(t) decreases, as could be expected. Note also, as a matter of course, that:

$$\int_T^{\infty} rKe^{-r(T-t)}dt = K$$

The Consumption of Oil and the Supply of Nonoil Energy Materials

The energy problem, at least in the short to medium run, is an oil problem. Not only is it almost impossible within the next decade to expand by a sizable amount the production of alternative forms of energy, but also even if it were possible, oil plays a unique roll in the energy picture in that in certain uses it is thought of as essential. Years of purchasing low-cost oil have given its consumers an almost unshakable confidence in its adaptability, and this can be seen in its low price elasticity. In the short run, the elasticity may be on the order of -0.1: A price rise of 1 percent causes a fall in demand of only 0.1 percent. By no means is this elasticity larger than -0.3, and just now this is probably true even in the medium run. (The price elasticity of gasoline in the United States appears to be even smaller.)

In this chapter we examine the demand for oil, with a large part of the discussion constructed around some of the better-known energy prognoses. Given the cost of these prognoses, it would be rude or ignorant to neglect them or to attempt to provide alternatives at present. Quite simply, they represent the best that money can buy; but as Ulph (1980) has indicated in an important survey article, collectively their shortcomings may equal their merits. In this chapter we also make a short survey of uranium, coal, and some other energy options, since the demand for oil may depend on the availability of these materials. Before we begin, some background to our forthcoming discussion can be supplied in numerical form (see table 5-1).

Just what these figures would have been without the oil-price shock of 1973-1974 cannot be speculated on here, but one thing seems evident: In the short run at least, the consumption of energy seemed scheduled to expand dramatically under the pressure of a growing prosperity in both Western and Eastern Europe. Remember that much of the demand for energy is *affluence-related,* to use a term of Fisher (1974). As more countries approach a North American standard of living, energy consumption would rise accordingly. More energy is used in cities than in rural districts, and with the increasing urbanization that would take place in the more affluent, less developed countries, energy intensities would rise sharply. Similarly, more energy is used in factories and offices than in the home, and had economic development continued in Western and Eastern Europe at the pre-1973 pace, larger numbers of women would have entered nondomestic employment. At the same time, as pointed out in Banks (1977), there is no reason to expect energy consumption in some of the mature industrial coun-

Table 5-1
Some Energy-Use Statistics for North America, Western Europe, and Japan

	North America	*Western Europe*	*Japan*
Yearly Growth Rate of Energy Production (%)			
1960-1970	4.5	—	-2.2
1970-1980	1.0	6.3	4.5
1974-1980	1.2	8.6	5.5
1960-1980	2.7	3.1	1.1
Yearly Growth Rate of Oil Imports (%)			
1960-1970	5.2	12.6	19.9
1970-1980	11.6	0.4	6.2
1974-1980	9.9	-2.1	4.8
1960-1980	8.4	6.3	12.9
Consumption of Various Types of Primary Energy 1976 (Mtoe)			
Coal	365.0	235.0	52
Hydroelectric	113	74	21
Nuclear	46	27	7
Natural gas	580	175	11
Oil	894	697	255
(Oil imports)	(354)	(607)	(254)
Total	1,998	1,208	346
Some End Uses of Oil (Mt)			
Industry	98	141	61
Transportation	428	158	38
Road transport	358	121	29

Source: OECD. *Energy Balances of OECD Countries*, 1978; *Energy Outlook to 1989*, 1977; BP Statistical Review of the World Oil Industry, 1978 & 1979.

tries to continue to grow at historic rates: A logistic evolution, which implies near saturation at some finite level of energy consumption per capita, seems more realistic, particularly since suburbanization and female employment outside the home cannot continue indefinitely. Still, if the rapid natural increase in the world's population and the change in lifestyles in the less developed world are given their proper weight, the overall prospect would be for the growth of energy demand to continue at its previous rate unless hampered by shortages of one type or another originating on the supply side. *Per capita* rates of energy use, however, may be slowed somewhat.

*Elasticities of Demand

This section is optional. It contains the sort of material found in many intermediate books on economic theory and is included here because of its overall importance for this chapter. All readers who can follow it are advised to do so, although as was made clear earlier, the remainder of this book can be read even if this material is skipped. A mere comprehension of elasticity will suffice, and this reduces to the following: The *price elasticity* is the percentage change in the demand evoked by a 1 percent change in price, with all else constant, while the *income elasticity* is the percentage change in demand caused by a 1 percent change in income (with all else constant).

The next step is to review the concept of elasticity in the context of the theory of demand. Thus, a demand curve shows the demand for a commodity, given its price. The thing to realize is that *market demand curves* are an aggregate of individual demand curves, and basically these individual curves need not resemble one another. Take, for instance, the situation shown in figure 5-1, where we have portions of two individual demand curves that we aggregate into a market-demand schedule.

The aggregation here is straightforward and takes place horizontally. The value of Q_2 in figure 5-1c is simply the sum of q_{12} and q_{22} in figure 5-1a

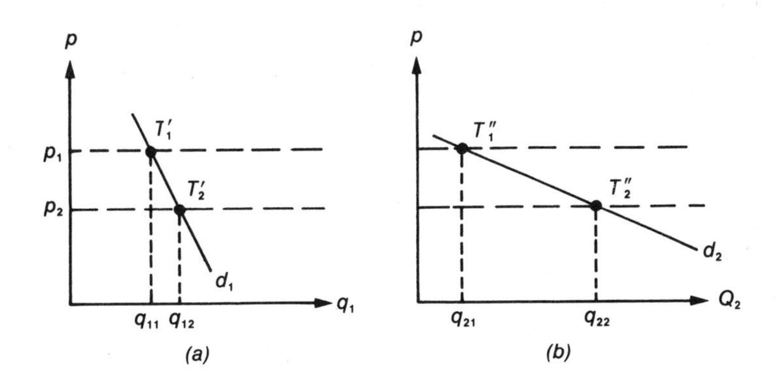

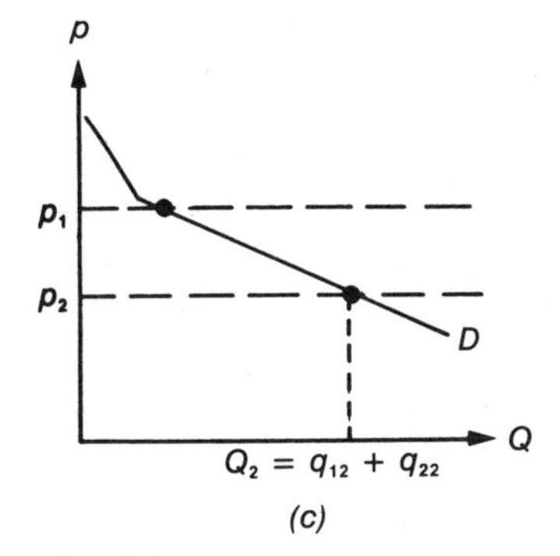

Figure 5-1. Typical Demand Curves and their Aggregation

and *b,* respectively, given a value of the price. What we should take particular care to notice are the elasticities of the individual demand curves. In figure 5-1*a* a decrease in price from p_1 to p_2 causes a relatively small increase in demand from q_{11} to q_{12}; in figure 5-1*b* the increase in demand, for the same price decrease, is much larger: $q_{22} - q_{21}$. For illustrative purposes, take $p_1 = 100$, $p_2 = 99$, $q_{11} = 1{,}900$, $q_{12} = 1{,}910$, $q_{21} = 1{,}900$, and $q_{22} = 2{,}000$. In this case we have a 1 percent *decrease* in price [(100 − 99)/100 × 100 percent] leading to a 0.526 percent [(1,910 − 1,900)/1,900 × 100 percent] increase in quantity in figure 5-1*a.*

We can now calculate one version of elasticity, the *arc* elasticity. In all cases elasticity is defined as the percentage change in quantity divided by the percentage change in price (or, the percentage change in quantity resulting from a 1 percent change in price). Since in the normal case these changes take place in opposite directions, the *price elasticity of demand* is defined as being negative. Thus in figure 5-1*a* the elasticity is $0.526/(-1) = -0.0526$. This is greater than -1, indicating that a 1 percent decrease in price is not compensated by an increase in demand, and thus total revenue pq falls. Specifically total revenue goes from $p_1q_{11} = 190{,}000$ to $p_2q_{12} = 189{,}090$. Over the stretch, or arc, on which the elasticity has been calculated, demand is *inelastic.* By way of contrast, in figure 5-1*b* elasticity is $5.26/(-1) = -5.26$, and the demand is *elastic.* The reader can check what this means for the change in revenue.

We now face a slight problem as a result of calculating the elasticity over an arc rather than at a point. If, instead of a price fall, we had a price increase, say from 99 to 100, then we would get a different value of the elasticity. The issue here can be clarified by writing the formula for the arc elasticity:

$$e = \frac{\Delta q/q}{\Delta p/p} = \frac{(q_{11} - q_{12})/q}{(p_1 - p_2)/p} = \frac{q_{11} - q_{12}}{p_1 - p_2}\left(\frac{p}{q}\right)$$

Notice that regardless of whether we have an increase (or a decrease) in the price equal to $p_1 - p_2$ (or $p_2 - p_1$), there is no change in the *absolute value* of $q_{11} - q_{12}$. It is p and q (or p/q) that change, depending on whether we have an increase or a decrease in price. For example, if the price decreases, we have $p/q = p_1/q_{11} = 100/1{,}900 = 0.0526$ in figure 5-1*a*. In the same figure, with an increase in price we would have $p/q = p_2/q_{12} = 99/1{,}910 = 0.0518$. This leads to different value of the elasticity over the same arc. At this point the reader should go through the same type of calculation employing the data given for figure 5-1*b*.

This discrepancy is handled by making p and q averages of the two prices and quantities. Thus arc elasticity e in figure 5-1a becomes

$$e = \frac{q_{11} - q_{12}}{p_1 - p_2} \frac{\frac{(p_1 + p_2)}{2}}{\frac{(q_{11} + q_{12})}{2}} = \frac{q_{11} - q_{12}}{p_1 - p_2} \frac{p_1 + p_2}{q_{11} + q_{12}}$$

Using the numerical values given, for the arc elasticity over $T'_1 - T'_2$ in figure 5-1*a*

$$e = \frac{-10}{1} \cdot \frac{199}{3{,}810} = -0.5223$$

The reader should now calculate the arc elasticity over $T''_1 - T''_2$ in-figure 5-1*b*. The following should also be observed. Elasticities often are expressed in absolute values. Thus an elasticity of -0.5223 is simply called 0.5223, with the understanding being that in the case of a normal demand curve, demand changes are in the opposite direction from price changes. The key thing to remember here is that an elasticity which is smaller than 1 [or greater than -1 if we are observing signs (for instance, 0.75 or -0.75] signifies that a given percentage change in price causes a smaller percentage change in quantity. The significance of all this for revenue is that when the demand is inelastic, a decrease in price means that even though demand might increase, it will not increase enough to prevent a fall in revenue. Analogously, when the demand is inelastic, an increase in price, means a percentage decrease in demand that is less than the price rise. Consider, for example, the situation with oil. Even when the price increased by almost 400 percent, demand fell only marginally, and producer revenues not only were maintained, but also increased drastically.

Before we leave this aspect of our subject, it should be observed that the most interesting problem involves starting out with the value of the elasticity and calculating changes in demand. For instance, if we assume that the (short-run) elasticity of oil is 0.1, then a 25 percent increase in price with all else constant means a percentage change in quantity of $\Delta q/q = n_p(\Delta p/p) = -0.1\ (25) = -2.5$ percent, where n_p is the price elasticity of demand.

Next we consider the income elasticity of demand. This concept is similar to the one just examined, and we are interested in the percentage change in demand, given a certain percentage change in income. Calling the income elasticity of demand n_i and income *I,* we have

$$\frac{\Delta q/q}{\Delta I/I} = n_i$$

Figure 5-2*a* shows a typical shift in the demand curve under the in-

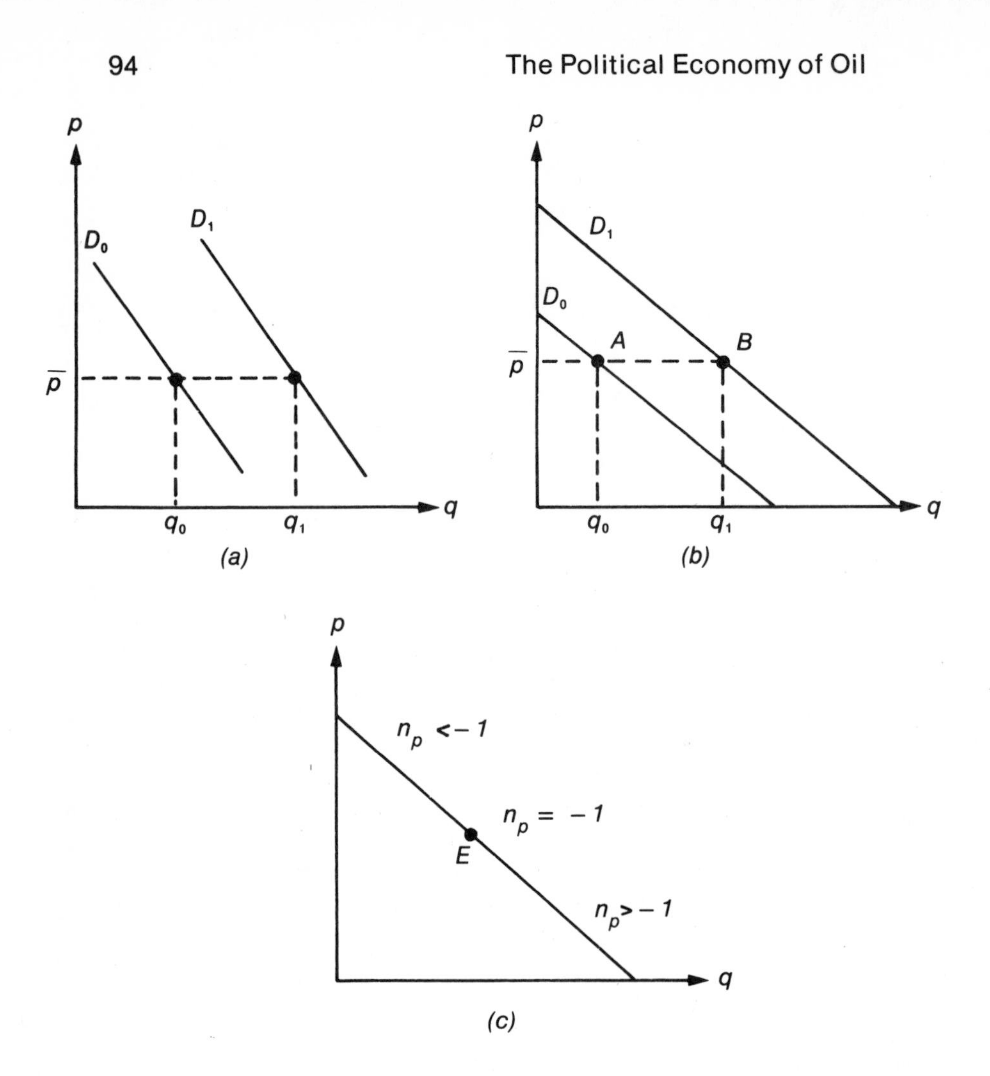

Figure 5-2. Elasticities on Linear Demand Curves

fluence of an increase in income: With the price constant at $\overline{p}$, demand increases from q_0 to q_1 when income increases from I_0 to I_1. For instance, suppose we have an income elasticity of demand of $n_i = 1.1$ and $I_0 = 100$ and $I_1 = 150$. We thus get $\Delta q/q = n_i (\Delta I/I) = 1.1(50/100) = 0.55$. Here income increased by 50 percent, whereas demand increased by 55 percent. In this example $I = I_1 - I_0 = 50$.

We can look at the effect of a parallel shift to the right of a linear demand curve on the price elasticity of demand. Using the definition of price elasticity given above, we note that, with an unchanged price, the absolute value of the price elasticity at B in figure 5-2b is smaller than that at A. This is so since, with $n_p = \frac{(\Delta q/q)}{(\Delta p/p)}$, at B the only change is in q, which is larger: $\Delta q/\Delta p$ is the same for D_0 and D_1 and, as mentioned, p, is constant and equal to $\bar{p}$. [The expression *absolute value* is important. If n_p equals -0.3 at A and -0.2 at B, then the elasticity is actually larger at B if we concern ourselves with the sign; but the absolute value (the value considered without the sign) is smaller.] If the reader has any doubts about this matter, he or she should construct a numerical example or two, since this is a significant topic. The short-run price elasticity of oil is very low, while its medium- to long-run elasticity (as defined below) is somewhat larger. But because the demand curve for oil is shifting to the right under the effect of an increase in income in the oil-importing countries, the fall in demand that eventually would be induced by the major price rises which have taken place recently is moderated to a considerable extent.

We continue by noting that on a linear demand curve, the elasticity changes as we move along the curve. As the reader can easily check, in figure 5-2*c* the demand curve is elastic above point *E*: A decrease in price, for example, would lead to a percentage decrease in demand larger than the percentage decrease in price. However, the curve is inelastic below point *E*.

The change in elasticity as we move along the demand curve is one reason why curves with a constant elasticity of demand are popular in economics. These have the appearance shown in figure 5-3a and resemble a hyperbola in that they never meet the p or q axis.

The most important reason for using this type of curve, however, is that it gives us a chance to put the concept of short- and long-run elasticities in a usable analytic framework. Take a situation of the type shown in figure 5-3*b*, where the price suddenly rises from p_0 to p_1. In the short run there is only a small decrease in demand, from q_0 to q_1, as the consumer moves up a (constant-elasticity) short-run curve from W to X. In the long run, however, some adjustment to the new high price may be possible in the sense that more of the commodity can be dispensed with. For example, if the commodity is fuel, in the medium to long run, vehicles that economize on fuel can be introduced. Thus demand would continue moving toward q_2, which is on the long-run demand curve D_L. In terms of figure 5-3*b*, this movement can be depicted as going from X to Y; a large number of demand curves are being crossed, one of which is dashed and designated D_m (where M signifies intermediate). Another way of showing the movement of price and quantity can be found in figure 5-4*a*, where the change in price is shown in step form, and in figure 5-4*b*, where we observe an asymptotic movement to the new equilibrium q_2.

The situation with the elasticities is portrayed in figure 5-4*c*. Here, in order to avoid using negative numbers, absolute values are employed. For instance, as elasticity moves from n_S on curve D_S to n_L on curve D_L in figure 5-4*b*, where n_S and n_L are negative; in figure 5-4*c* these elasticities

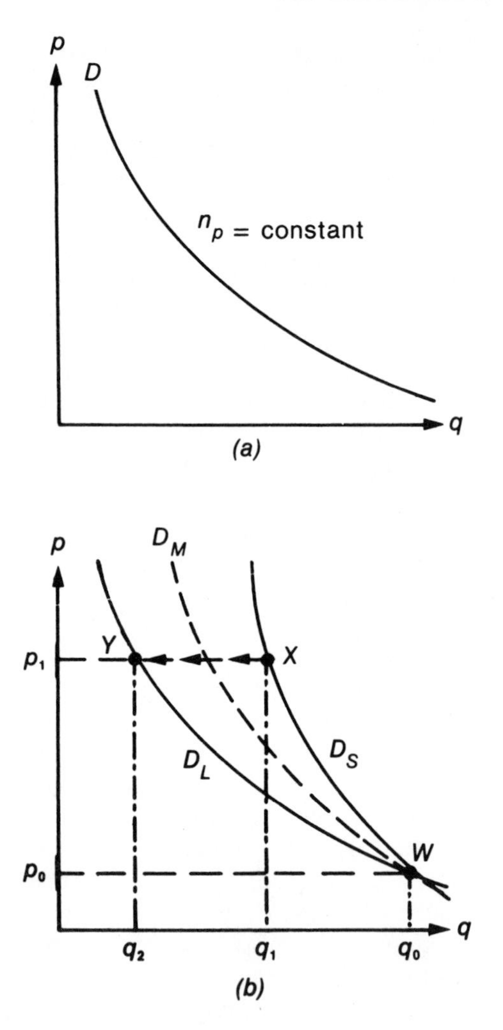

Figure 5-3. Constant-Elasticity Demand Curves and the Concept of Short- and Long-Run Elasticities

are shown in absolute terms (that is, they are positive) and are designated $|n_S|$ and $|n_L|$. As we move from *X* to *Y* in figure 5-3*b*, we are crossing demand curves whose elasticities, in absolute terms, are between $|n_S|$ and $|n_L|$. Finally, let us reemphasize that under the pressure of increased income, a system such as the one shown in figure 5-3*b* is in motion to the right. Thus if we take the case of oil, where the absolute value of the short-run elasticity is low, when the price is raised by a large amount, demand

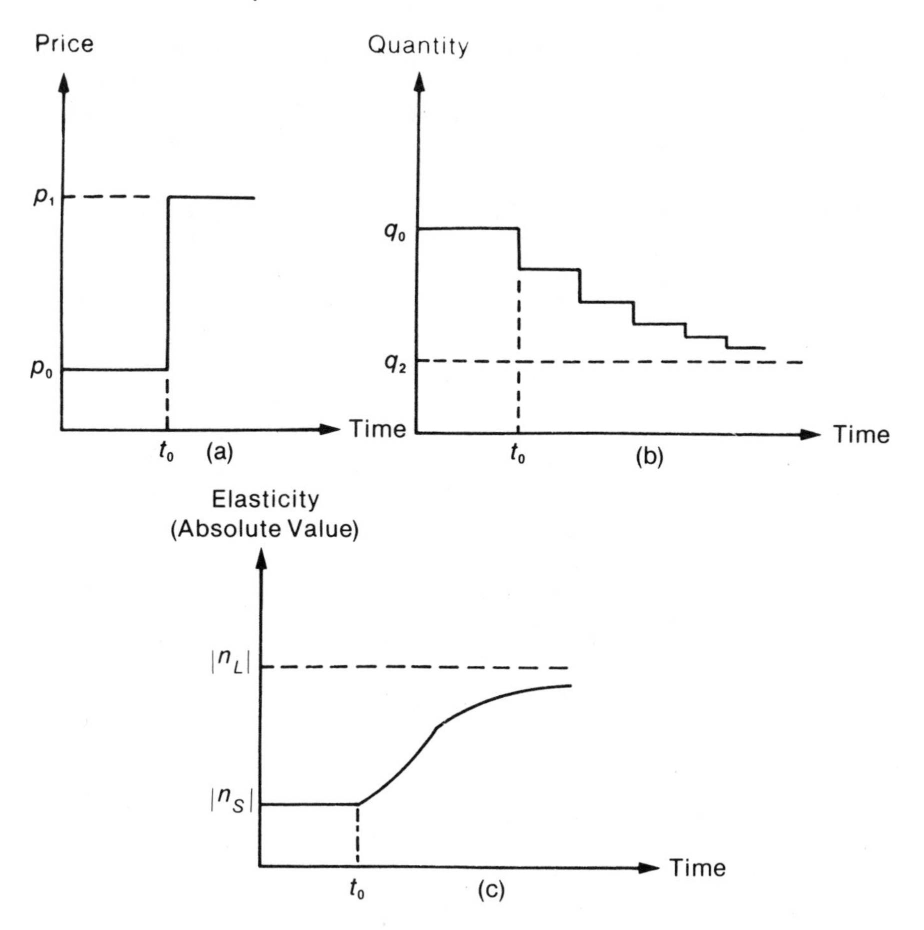

Figure 5-4. Movements to a Long-Run Equilibrium Following an Increase in Price

falls by only a small amount, (although in the medium to long run there are possibilities for a major fall in the demand if the income remains constant). But, as we said, the income does not remain constant—it increases and thereby substantially reduces the fall in demand. In figure 5-3*b*, for example, demand would not reach q_2 because the movement from q_1 to the vicinity of q_2 would take considerable time; and meanwhile the long-run demand curve would have shifted to the right.

The Consumption of Oil

We now consider some projections of the consumption of oil. The interesting aspects of this topic are not the projections themselves (since the recent oil price rises and the new tendency toward conservation in some of the most important OPEC countries have changed the background for these projections in such a way as to make their validity highly dubious), but the issues they attempt to come to grips with. The best-known prognosis of the future energy situation is that of the International Energy Agency (IEA) of the OECD, which was published in the report *World Energy Outlook* (WEO). Four scenarios were constructed for the purpose of judging oil consumption in 1985, and these employed yearly growth rates of from 3.7 to 4.7 percent. The real price of oil was assumed to be unchanged over this period, and no price elasticity was specified for these calculations. However, the average energy-GNP elasticity was assumed to be falling over this interval. This, of course, is a natural consequence of a marginal energy-income or energy-GNP elasticity that is less than unity. With the assumed rates of economic growth, energy use could fall by about 5 percent during a ten-year period.

As indicated in table 1-2, the demand for OPEC oil was determined as a residual of the total energy demand. Among other things, this means that errors in the calculation of the availability of other energy resources could result in serious discrepancies in this residual entry. There is also the problem of estimating the energy demand of the LDCs as well as the entire energy supply and demand in the centrally planned countries, particularly the USSR. At present, certain observers tend to believe that Russia will be a net importer of energy in a few years instead of the net exporter of today. I believe, however, that given Russia's domestic supplies of gas and coal and willingness to build and use nuclear installations, there is scarcely any economic reason for the Russians to import oil if its price continues to rise at its present rate. On the contrary, a more reasonable assumption is that Russia will make every effort to continue exporting this commodity.

In the OECD report there is no extensive discussion of OPEC behavior in the 1980s; and since work on this report began during a period when it was fashionable to be optimistic about the supply proclivities of major producers such as Saudi Arabia and Iran, it could contain some serious biases. Among its more notable passages is the one containing the prediction that OPEC will have a capacity of 45 Mbbl/d by the end of 1985, but there is no indication as to whether the actual production of that organization will attain this level. I believe, though, that it would be prudent to think in slightly less generous terms. Table 5-2 summarizes the OECD estimates of the demand for OPEC oil in 1985.

The Workshop on Alternative Energy Strategies (WAES) Study

This project, which brought together energy analysts from a number of countries, was primarily concerned with examining energy demand and sup-

Table 5-2
OECD Forecasts of the OECD Demand for OPEC Oil in 1985

GNP Growth[b]	*Income (GNP) Elasticity*	*Oil Price ($/bbl)*[a]	*Demand (Mbbl/d)*	*Conservation Policy*
High	0.84	11.51	47.0	Existing
Reference	0.84	11.51	39.3	Existing
Low	0.84	11.51	33.3	Existing
Reference	0.84	11.51	28.6	Vigorous

Source: OECD. Energy Outlook to 1985, 1977.

Note: No price elasticity of demand has been assumed, and the mean value for the total energy demand of the OECD is 102 Mbbl/d.

[a]This is price per barrel in 1975 dollars (the real price of oil is assumed to remain at the 1975 level).

[b]Reference growth rate 4.3 percent, low 3.7 and high 4.7 percent.

ply projections at regional levels, paying particular attention to *excess* supply and demand situations which might arise given various assumptions about the price of oil and the rate of economic growth. Attention was then paid to precisely how these regional disparities could be eliminated, which means that all country and/or regional projections were brought together and various patterns of imports and exports postulated and examined.

Six scenarios were eventually chosen for final study, of which two concerned projections up to 1985. In these scenarios there were two levels of economic growth and three possible oil prices: $17.25/bbl, $11.50/bbl, and $7.66/bbl, where these are assumed to be the prices in 1985. There were also some assumptions about energy policies in the main consuming countries. A price elasticity of demand for oil of −0.3 was used, and energy-GNP elasticities of 0.95 for OECD countries and 1.1 for other areas were assumed. As for the income elasticity for oil, this was taken to be 0.8 to 0.9, depending on the area. The total demand for oil in 1985 with the high-growth assumption was eventually calculated to be 63 Mbbl/d, while the total demand for oil with the low-growth assumption was 58 million barrels per day (Mbbl/d). Of these two demands OPEC was pictured as supplying 38.3 Mbbl/d in the first case and 36 Mbbl/d in the second.

Some criticism has been leveled against the WAES report. In his important survey of energy models Ulph (1980) notes that such a wide range of techniques have been employed by the different country teams responsible for the various components of the report as to bring into question the consistency of the integrated forecast. Unfortunately I cannot agree with this judgment. The question here is one not of integrating mathematical models but of reconciling numerical estimates prepared by individuals with firsthand experience of local supply-and-demand situations. No known econometric techniques are capable, in practice, of carrying out this assignment, and I predict that none will be developed in this century. However, it is quite conceivable that a programming model capable of designating the optimal pattern of world energy flows could be constructed. The problem is that optimal and actual patterns are usually entirely different things, given the way in which the real world functions. In fact, I posit that the strength of the WAES and similar studies is that they were not overly concerned with equilibrium situations and thus were not put in the position of having to introduce ad hoc simplifications in order to arrive at textbook solutions of textbook issues. As far as I can tell, the value of theoretical models of the Nordhaus and Manne (1976) type is found exclusively in the questions they raised. As was argued in chapter 4, the Nordhaus construction (which was filled with interesting and important theoretical points) provided an essentially false picture of short-term energy availability and contributed to the insane hankering after an OPEC collapse which ostensibly would lead to a collapse in the oil price. Here it should be noted that the WAES low-price scenario, which still hypothesizes an oil price that many "experts" considered unrealistically high both before and after September 1973, resulted in a huge gap between demand and supply in all the important energy forms. Nordhaus came to the naïve conclusion that the "efficiency" price of oil was $3/barrel in 1975. Thus, using his logic, the price of oil today should be less than $6/barrel.

The CIA, OPEC, and Exxon Forecasts

Like Professor Friedman, the CIA has a penchant for being wrong where certain important issues are concerned, but seems to have learned how the international oil market functions. To begin, the CIA visualizes the United States as being able to supply only a decreasing fraction of domestic requirements. The output of the Alaskan North Slope has already reached a plateau, the offshore U.S. area has been a disappointment, and in the last decade additions to proved reserves have been only half as large as production. In Western Europe production will more than double in the next few years, from 1.8 to 3.9 Mbbl/d, thanks to the North Sea oil fields, but North Sea production should peak at 3.5 Mbbl/d soon and decline afterward.

In the Third World some new sources of production are coming onstream in Mexico, Egypt, India, and Malaysia. These are expected to yield about 7 Mbbl/d in the mid-1980s, with Mexico contributing approximately 2.5 Mbbl/d. Balancing this production, however, will be a growing demand from the LDCs. The CIA also believes that Soviet oil output will peak soon and that oil requirements in Eastern Europe will remain high. Where China is concerned, the CIA does not believe it to possess the large amount of exploitable reserves with which it is often credited by the popular press. For the socialist bloc as a whole, the prediction is that its present export surplus will be transformed into an import surplus.

A generally pessimistic view is taken of OPEC's future willingness to export energy resources. The "preferred" output of Saudi Arabia is put at 8.5 Mbbl/d, with anything over this amount regarded as a concession to the West (that is, the United States). An Iranian output of 4 to 5 Mbbl/d by 1982 is seen as possible but not probable, while it is claimed that the oil minister of Iraq would like to see its output reduced from its present level of 3.5 Mbbl/d to about 60 percent of its long-term capacity, or 2.4 Mbbl/d. Both Kuwait and Abu Dhabi have recently come out for conservation, and at least Kuwait may have already started putting this predilection into practice.

The methodological approach of the CIA has been to assume some macroeconomic growth rates and income-energy elasticities and then to calculate the energy demand of various regions. The OECD domestic energy production is calculated, as is the energy supply of the other regions. In light of these calculations, the overall pattern of surpluses and deficits is determined, and a likely pattern of exports and imports is postulated. In the 1977 CIA prognosis this technique, employing a growth of income in the OECD of 4 percent annually, led to the conclusion that Saudi Arabia's production capacity would have to be increased to 23 Mbbl/d if OPEC were to export the 47 to 51 Mbbl/d that would make the CIA's assessment consistent. At that time the CIA failed to detect some of the responses to this figure emanating from the direction of the Arabian Gulf. Its latest forecast is much more in tune with reality: If gross national product in the OECD grows at only 3 percent a year, a major escalation in the oil price is unavoidable from about 1982 to 1983. The CIA's 1979 predictions are shown in table 5-3.

Finally we come to the prognoses of OPEC and Exxon. As was pointed out be Björk (1978), OPEC's calculations are an amalgam of a number of other studies. On the basis of a yearly world population growth of 2 percent and a rate of growth of 3.5 percent in the United States, a rate of increase of 4.6 percent annually is arrived at for the energy demand, although it is claimed that vigorous savings measures could reduce this to 3 percent. Nuclear energy is seen as providing an equivalent of 11 Mbbl/d of oil by 1985, which is more than twice that predicted by any other source. Coal also plays a role in the 1985 energy picture that is considerably larger than in the other prognoses. Still the demand for oil in 1985 is 55 Mbbl/d, with a 12-Mbbl/d supply originating in the United States, a 6-Mbbl/d supply from Western Europe, and a 2-Mbbl/d supply from Canada. In contrast, the WEO places the combined OECD output in 1985 at 17 Mbbl/d. The total demand for OPEC resources by the OECD is then 35 Mbbl/d, and with OPEC's internal demand plus the demand of the LDCs, OPEC will have to produce 40 Mbbl/d. As an aside, it is given that this is about 70 percent of the 57-Mbbl/d capacity which OPEC is presumed to dispose of at that time (as compared to the 45-Mbbl/d capacity forecast by the WEO).

Table 5-3
OECD Energy Supplies in 1978, 1979, and 1982 (Estimated)
(Mbbl of oil per day equivalent)

	1978	*1979*	*1982 (Estimate)*
Oil			
Production	13.7	14.3	14.8
Net imports	25.7	25.7	23.1
Total[a]	39.7	40.0	37.9
Natural Gas			
Production and imports	14.9	14.9	15.5
Coal			
Production and Imports	13.2	14.2	15.4
Nuclear	2.6	3.0	4.7
Hydro, Geothermal, etc.	4.8	5.2	5.8
Total	78.2	77.3	79.3

Source: U.S. Central Intelligence Agency, "The World Oil Market in the Years Ahead" (Washington, 1979).
[a]Including stock decreases.

Exxon's forecast is predicated on an energy growth rate of 3.9 percent per year for the non-centrally planned world. The total demand for OPEC oil in 1985 comes to 45 Mbbl/d, which would mean that capacity for a production of at least 50 Mbbl/d would have to be available. Explaining these figures is simple: They were calculated in 1977. Were Exxon making it projections today, it would have a somewhat less sanguine view of OPEC's output in the next few years. In fact, at the last world energy conference nobody expected OPEC production in the near future to increase much beyond the 31 million barrels per day produced in 1979.

Uranium

The principal use for uranium is as a fuel in the generation of electricity. The future demand for uranium thus will be tied to the rate of growth of the demand for electricity, as well as to the price of oil, since the oil-fired power station has generally been the most important supplier of electricity for the last half-century.

The consumption of electricity has been increasing at a rate of 7 to 8 percent a year, which means that electricity consumption doubles about every ten years. The World Energy Conference in 1977 estimated that in the year 2020 about six or seven times as much electricity will be required as at present and that electricity generation will use 40 percent of all primary energy. Mackay (1978) estimates that electricity demand will increase by at least 6 percent per year until the end of this century and probably well into the next. Furthermore, where the primary energy sources used in electrical generation are concerned, nuclear power will increase from its present 5 to 6 percent to 45 percent of the total. Until the recent oil-price rise, much of this gain was scheduled to be at the expense of coal, but whether this will actually be the case now is uncertain. The Atomic Industrial Forum and the United Kingdom Atomic Energy Authority in 1977 claimed that nuclear energy is more economical for generating electricity than other fuels, if we assume that all else remains the same. Their figures were as follows: For the United States in 1977, the cost in cents per kilowatthour for electricity generation was 1.5 for nuclear, 1.8 for coal, and 3.5 for oil; for the United Kingdom, in pence per kilowatthour, 0.69 for nuclear, 1.07 for coal, and 1.27 for oil. But the State Electricity Commission of Victoria, in its evaluation of the Loy Yang project, calculated that a power station employing inexpensive Australian brown coal was superior to a nuclear power station delivering the same amount of electricity, and similar contentions have been made elsewhere. To my way of thinking, this issue is still open, with the crucial factor being the availability of coal as well as the availability and expense of the equipment that will keep this coal from causing too much damage to the environment.

In the matter of uranium supply, it can be said that although until the mid-1970s predictions were that reserves of high-grade uranium ore with low extraction costs would soon be exhausted, a scaling down of the estimates of uranium demand, together with an increase in the estimates of reserves, should mean that supplies will be adequate for the next ten to fifteen years without a major price increase being necessary. The cumulative production of uranium up to 1976 was about 471,000 t, with most of this coming from North America and southern Africa. The present world production capacity is in excess of 40,000 tonnes per year (t/yr) of uranium, but the general belief is that present resources will support a production of 110,000 t/yr by 1990. If we exclude the USSR and China, total reserves of *reasonably ensured* low- to medium-cost uranium are estimated at 2 to 2.1 Mt and about the same amount in the estimated-additional-resources category. In case the reader is interested in what 2 Mt of uranium amounts to in energy terms, with all constraints in the nuclear fuel cycle taken into consideration, it would be equivalent to about 95 trillion barrels (Tbbl) of oil. About three-fourths of the world's uranium supplies are located in Australia, Canada, South Africa, Sweden, and the United States. As for the expression *low to medium cost,* at present this signifies a cost for uranium concentrate (U_3O_8) of less than $50/lb, where 1 lb of uranium metal is equal to 1.8 lb of U_3O_8.

Having made, in brief, some of the most important points in this section, we can now systematize our discussion. To begin, if we look at the market for uranium, we note in the beginning a very strong military demand emanating from the nuclear weapons programs of the various superpowers. In the United States this meant the regulating of the market for a long time by the Atomic Energy Commission (which also dominated the buying side of the uranium market for almost twenty years). The big demand for uranium to use in civilian projects began early in 1962, and since that time civilian demand has been the largest component of the market.

There are six principal stages in the manufacture of nuclear fuel, of which we are mainly concerned with the first two: the mining of uranium ore which, like coal, is done either underground or in open pits and processing the ore to obtain uranium oxide (U_3O_8), also known as yellowcake. The uranium ore mined today has, on average, a uranium content of between 0.15 and 0.30 percent. The Canadian mines providing the supply of uranium to the United States in the mid-1940s mined ore with a uranium content of up to 1 percent; but few deposits having this richness have been located in the past forty years, and there is genuine concern that toward the end of this century, good-quality ore will start becoming scarce. (Sweden, for instance, has large supplies of ore, but at present these are in the high-cost category, with uranium contents in the 0.02 to 0.03 percent range.) Should it appear that a shortage of ore is imminent and should the demand for uranium be accelerating at that time because of a speedup in the construction of nuclear plants, then there may be an irresistible pressure on governments to introduce the breeder reactor, which by breeding plutonium would increase supplies of uranium by a factor of up to sixty. For instance, Britain may already have enough plutonium and uranium 238 in the form of tailings from enrichment plants to be energy-independent for several decades if there is a switch to a breeder-based energy technology.

A shortage of uranium may also lead to a more intensive mining of less rich ores. As indicated above, even high-cost uranium would not greatly increase operating costs for nuclear plants. There is no shortage of claims that a tripling or quadrupling of the price of uranium would lead to the multiplying of exploitable reserves by a factor of five or ten. Here, though, it should be noted that the mining of extremely thin ores could cause environmental damage of a type corresponding to the more extreme ravages of strip-mining, and almost as much energy would have to be expended in mining and processing uranium as would be extracted from it in a light-water reactor. Thus the authorities might feel an additional compulsion to introduce the breeder.

Next we look at some figures for uranium reserves, production, and anticipated production (see table 5.4). It is interesting to note that in Australia exploration has slowed down because of a large-scale opposition to uranium mining *and* nuclear energy, although Australia may have between 20 and 30 percent of the world's uranium resources. Canada may have even more, but it is hardly economical to intensify the search for them at present. Just now Candian production is concentrated around Elliot Lake in Ontario, and

Table 5-4
Uranium Reserves and Resources and Attainable Uranium Production and Capacities

	Reasonably Ensured Reserves at $30/lb ($10^3$ t)	*Estimated Additional Resources at $30/lb ($10^3$ t)*	*Reasonably Ensured Reserves at $30/lb to $50/lb ($10^3$ t)*	*Estimated Additional Resources of $30/lb to $50/lb ($10^3$ t)*	*1977 Total*	*1977 Production*	*1977 Capacity*	*OECD Production[a]*	
								1985	*1990*
Australia	289	44	7	5	345	0.4	0.4	11.8	20.0
Canada	167	392	15	264	838	6.1	6.1	12.5	11.2
United States	306	838	42	215	1,401	11.2	14.7	36.0	47.0
Republic of South Africa	523	34	120	38	715	6.7	6.7	—	12.0
Niger	160	53	—	—	213	1.6	1.6	9.0	9.0
Gabon	—	—	—	—	—	—	0.8	1.2	1.2
Other Africa	58	65	6	10	139	—	—	—	—
France	37	24	15	20	96	2.2	2.2	3.7	4.7
Spain	7	8	—	—	15	0.2	—	1.3	1.3
Sweden	1	3	300	—	304	—	—	—	—
Argentina	18	—	—	—	18	—	—	—	—
Brazil	18	8	24	—	42	—	—	—	—
Other Latin America	5	7	—	1	13	—	—	—	—
Other Western Europe	21	10	12	25	68	—	—	—	—
India	30	24	—	—	54	—	—	—	—
Japan	8	—	—	—	8		—	—	
Other	—	—	3	7	10	—	—	—	—
Total	1,647	1,510	544	585	4,287	28.6	33.0	92.0	110

Source: World Energy Conference Conservation Commission, 1978; OECD, Energy Prospects to 1985; Workshop on Alternative Energy Strategies, 1977.

[a]Estimated by OECD (1977). In 10^3 *t*/yr of uranium.

most production is exported to the United States and the United Kingdom. Similarly, although Sweden may be proceeding with a nuclear program, no plans exist to start producing uranium, although some very small projects could begin during the 1980s. In the United States, uranium production began a long decline in 1960, but started to rise again in 1966, although it was not until 1976 that the 1960 output level was reached. United States production is located mainly in New Mexico, Colorado, Wyoming, and Utah, and there is some evidence that a large portion of the best ore in those states has been expended. Finally, in South Africa much uranium is won as a by-product of gold mining, and although it has been suggested that a large part of the South African supply of uranium is dependent on the price of gold, the quantity of gold produced per year in South Africa does not appear to depend on its price.

Our next topic is the demand for uranium. Here the key variable is the number of nuclear installations planned. The price of uranium is not particularly important, since the cost of uranium per period is only a very small percentage of the unit cost of electricity produced in a nuclear installation. In looking at the history of the demand for nuclear installations, we see a sharp downward adjustment in the mid-1970s resulting from opposition to nuclear energy on the part of enough people to make it politically unsatisfactory, an occasional stabilizing of the price of oil, and (to a lesser extent) uncertainty about business conditions. However, the present series of oil price rises is undoubtedly altering this trend. To get some idea of the nuclear picture just now and how it will probably appear in the near future, see table 5-5, which deals with nuclear capacity installed and on order at the beginning of 1979.

The unit of measure of capacity in this table is 1 million watts (1 MW), although the usual unit of measurement is 1 billion watts (1 gigawatt, or 1 GW). Also at least five different nonbreeder reactor types are being operated commercially at present; and the French, British, and Russians have constructed, and are operating, breeder reactors. Currently it is felt that the light-water reactor will remain the most important reactor category until the end of the twentieth century, claiming 80 or 90 percent of nuclear capacity, and after that a greater receptivity to the breeder and other exotic pieces of equipment will be shown. This assumption, of course, is based on the continued avoidance of serious nuclear accidents or incidents; however, I think that it is clear that a few more episodes of the Harrisbury, Pennsylvania (Three Mile Island), type would call for a thorough reestimation of future nuclear popularity. It has been claimed that the Harrisburg accident has already impinged on the demand for nuclear power in the United States: at the beginning of 1979 the U.S. Department of Energy estimated that 114,000 MW of nuclear power would be available by 1985 and 152,000 MW by 1990. The most recent projections are 95,000 MW in 1985 and 129,000 MW in 1990.

With this material as a background, it is fairly easy to evaluate the ap-

Table 5-5
Nuclear Power Plants: Existing, Under Construction, and on Order as of January 1979

	Installed		*Under Construction or on Order*	
Country	*Number*	*Capacity (MW)*	*Number*	*Capacity (MW)*
Argentina	1	319	1	600
Bangladesh	—	—	1	200
Belgium	4	1,660	4	3,800
Brazil	—	—	3	3,116
Bulgaria	2	840	2	840
Egypt	—	—	1	622
Philippines	—	—	2	1,240
Finland	2	1,080	2	1,080
France	16	8,321	32	32,195
Germany (FR)	15	8,857	15	16,436
India	3	580	5	1,080
Iran	—	—	6	7,192
Italy	4	1,447	5	3,959
Japan	23	14,523	8	6,857
Yugoslavia	—	—	1	632
Canada	11	5,516	15	10,056
Korea (South)	1	564	4	3,134
Cuba	—	—	1	420
Luxemburg	—	—	1	1,247
Mexico	—	—	2	1,308
Netherlands	2	500	—	—
Pakistan	1	125	—	—
Poland	—	—	—	—
Rumania	—	—	2	1,020
Switzerland	4	1,926	3	3,007
USSR	28	9,820	32	30,685
Spain	3	1,073	14	13,158
United Kingdom	18	8,118	6	3,750
Sweden	6	3,740	6	5,720
Republic of South Africa	—	—	2	1,844
Taiwan	2	1,220	4	3,714
Czechoslovakia	2	530	3	1,260
Hungary	—	—	4	1,680
United States	72	52,477	133	148,065
Austria	—	—	1	692
East Germany	4	1,340	7	3,780
Total	224	124,586	332	315,259

Source: Australian Atomic Energy Commission, 1978.

proximate forecasts of uranium requirements up to the year 2000, with emphasis on the word *approximate.* In 1977 a conference in Salzburg was informed by the IEA that between 1977 and 2000 between 1.65 and 3.6 Mt of uranium was necessary if existing plans for reactor construction were to be fulfilled; for the following year estimates amounting to one-half of these were being freely bandied about by the same organization. As far as I can tell, one of the most important factors determining uranium demand after 1990 will be a realization that the demand for oil is in the process of outrunning the supply. Under these circumstances, almost all upper limits on present forecasts could be surpassed, and a serious problem might arise in finding enough uranium to satisfy requirements for the following twenty years without a mass introduction of the breeder. Certain aspects of this situation have already been broached, particularly in France, where President d'Estaing insists that if uranium from French soil is used in fast-breeder reactors, French energy resources could be considered comparable to those of Saudi Arabia; in addition, his government claims that nuclear-generated electricity is about 45 percent cheaper than electricity generated by oil-fired plants. We can now examine some forecasts of the demand for uranium (see table 5-6).

We conclude this section with some comments on the price of uranium. Experts who have committed themselves seem to have widely varying expectations of the uranium price in 1985. The four individuals or organizations quoted in the Electric Power Research Institute report give (in 1975-1976 dollars) 47, 40, 40, and 55 to 77. Forward-market quotations for deliveries in the 1980s average about \$50/lb, but these quotations are apparently in *current* dollars, which would indicate an expected price of under \$40/lb in 1975-1976 dollars. One of the key factors determining price is, or course,

Table 5-6
Two Prognoses of the Demand for Uranium (in '000 of Short Tons of U_3O_8)
(10^3 tons of U_3O_8)

	EPRI[a] *Middle Case*	*OECD-IEA 1977 High and Low Alternatives*	*OECD-IEA Estimated Growth Rates (%/yr.)*
Demand during the year			
1980	40	50-60	—
1985	90	80-110	—
1990	140	110-200	—
1995	210	140-300	—
2000	270	160-440	—
Cumulative Consumption			
1975-1980	200	220	—
1975-1985	530	580-670	14-17
1975-1990	1,120	1,080-1,490	—
1975-1995	2,010	1,710-2,820	—
1975-2000	3,220	2,470-4,720	4.5-9.0

Source: OECD *Uranium, Resources and Production*, 1977; Electric Power Research Institute (EPRI), 1977.

[a]Electric Power Research Institute.

[b]The OECD-IEA high-low alternatives cover an accelerated nuclear scenario and a present-trend scenario.

Table 5-7
Production Costs for Uranium Concentrate for the United States in 1968[a]
($/lb)

Ore Grade (% Uranium)	*0.15*		*0.20*		*0.25*		*0.30*	
Ore (t/d)	*Underground*	*Open Pit*	*Underground*	*Open Pit*	*Underground*	*Open Pit*	*Underground*	*Open Pit*
500	11.09	7.54	8.75	5.99	7.30	5.10	6.34	4.50
1,000	9.35	6.51	7.45	5.32	6.26	4.56	5.47	4.05
2,000	8.37	6.04	6.71	4.96	5.67	4.27	4.98	3.82
5,000	7.54	5.65	6.09	4.67	5.18	4.07	4.57	3.62

Source: U.S. Bureau of Mines.
[a]Between 1968 and 1975 the amount of uranium obtained from open-pit mining in the United States increased from 35 to 55 percent; at the same time, the scale of operations increased by about 60 percent on average. Both factors have tended to decrease production costs.

costs, and these are shown for the United States in table 5.7. This table is for the year 1968, the last year, to my knowledge, for which such a comprehensive set of figures was available. An updating of these figures is a simple matter, however, if we assume that they have grown at the average rate of inflation in that country.

The uranium market also features a kind of cartel. The uranium cartel is called SERU (Societé d'Etudes de Recherches d'Uranium), and it came into existence in 1972. It is principally an organ for cooperation among Canada, Australia, South Africa, and France; its purpose is to establish common market prices and to divide up the market. Initially some minimum forward prices were established which were entered on contracts for delivery up to five years in the future, but with the upswing in U.S. prices that began in 1974, these contract prices were soon adjusted upward. Everything considered, it is impossible to say just what effect the cartel has managed to have on raising prices. At one time the U.S. market formed a separate entity because of import restrictions, but when these were removed in 1973 and the growing U.S. demand reached suppliers abroad, the price was quick to escalate upward.

Between 1965 and 1972 the price of yellowcake (U_3O_8) varied between $5/lb and about $8/lb. But in 1972 demand started to increase, and the price reacted accordingly. Then in 1973 the (U.S.) Atomic Energy Commission specified that contracts for uranium enrichment had to be placed eight years in advance. Many utility companies placed these contracts without bothering to secure feedstocks of uranium: They felt that since the price of uranium had been low for years, there would be no problem in obtaining desired supplies in the future. But they soon saw they were wrong, since during the period of low prices uranium producers had not expanded capacity. Thus, when they began to buy uranium, they found themselves paying prices that were much higher than anticipated. In 1976 the spot price of uranium reached $41/lb, and in general it has stayed around this level, although it reached $43/lb in 1978. Forward contracts have also been resorted to extensively in the uranium market. At this point the reader should peruse chapter 4 for a short exposition of forward contracts, because in this market they are written for a fixed *floor* (that is, minimum) price, plus a premium based on the spot price at the time of delivery. Figure 5-5 shows the price of uranium in dollars per pound of U_3O_8 in the United States since 1968.

Among the firms surprised by the price upswing mentioned earlier was Westinghouse of the United States. In 1972 and 1973 it was a very successful seller of nuclear power plants, for which it also agreed to supply U_3O_8 at $21/lb. But in 1974 the long-term uranium price doubled, and spot prices rose to $77 per kilogram ($77/kg). Westinghouse, from 1977-1981, needed

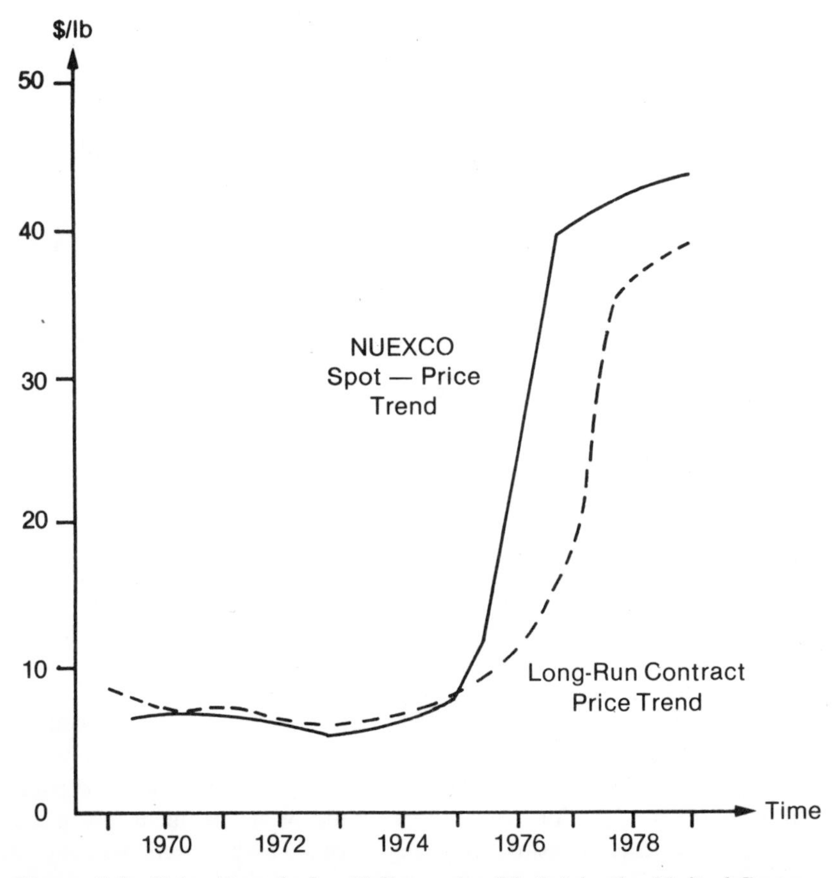

Figure 5-5. Price Trends for Yellowcake (U_3O_8) in the United States

16,500 t of uranium which, apparently, was available—but at four times the price Westinghouse considered that it could afford. This firm then announced that it could not fulfill its commitments, whereupon it was sued by various power companies, who in turn were countersued by Westinghouse. This case is still in litigation, but Westinghouse has claimed that it was the victim of an international cartel and has tried to bring charges against some of its Australian suppliers who, declining to appear in a U.S. court to which they had been summoned in January 1977, now run the risk of being arrested if they are so imprudent as to set foot in the United States.

Just now the spot price of uranium in the United States has fallen to $40/lb, and opinions have been expressed that the U.S. uranium industry is on the verge of a protracted spell of bad news. The opening of mines already projected are being delayed, and the exploration budget for this industry is expected to fall by 8 percent, to $314 million. The long-run result of this last phenomenon could conceivably lead to supply difficulties in the late 1980s, particularly if this trend deepens, since developing a uranium mine requires up to eight years and with the oil price capable of spiraling upward at any given time, there might be a strong boom in the demand for uranium before we are too many years into the 1980s. In the event of any domestic shortages, U.S. utility companies anticipate that Canada and Australia could provide them with uranium, and in fact several U.S. energy firms have begun exploration in Canada, where costs are lower and ore bodies are said to be richer. The situation with Australia remains uncertain, however, since although uranium production in that country is due to increase, it is impossible to say at which rate and exactly how much uranium Australia will be capable of exporting by the end of the 1980s.

Some figures have already been provided for energy costs. Here is a more complete roster of what it costs to produce 1kW of electricity in different countries using coal, oil, uranium, and water power. See table 5-8.

One of the most recent estimates I have seen contends that a 1,000-MW

Table 5-8
The Cost of Producing One kWh of Electricity in Eight Countries

Country	*Monetary Unit*	*Nuclear*	*Coal*	*Oil*	*Waterpower*
France	Centime	10.4	12.6	14.1	—
Denmark	Öre	—	40	40	—
England	Pence	1.02	1.29	1.31	—
Finland	Penni	10	—	—	15
Norway	Öre	—	—	—	6
USSR	Kopeck	0.6	—	—	0.15
United States	Cent	1.5	2.3	4.0	—
West Germany	Pfennig	7.8	11.5	—	—

Source: Australian Atomic Energy Commission; *The Petroleum Economist* (Various Issues). Bulletin of The U.N. Atomic Energy Agency, Vienna (Various Issues).

nuclear plant, in the United States, providing an electricity output equivalent to 10,000 bbl/d of oil requires an investment of $1.2 billion. It has also been suggested that if costs continue to increase at the present rate, by the end of the century the investment needed to provide nuclear facilities to replace 1 Mbbl/d could come to $120 billion.

Coal

After oil, coal is the second most important energy resource in the world. Between 2 and 2.5 Gtons/yr of coal has been used in the world during the last three decades, and in the 1970s coal was responsible for between 20 and 30 percent of the world's energy supply. The reason is clear. At 1978 oil prices, a coal power plant in the United States with full desulfurization facilities could produce electricity at an average total cost of up to 30 percent less than an oil-fired plant of comparable capacity.

Globally, the growth rate of coal consumption came to about 2.5 percent per year from 1860 to 1975. In the years between 1950 and 1975 it was 2.1 percent per year, but only 1.5 percent in the period 1965-1975. Some analysts believe that coal will provide as much as 33 percent of the world's primary energy in the year 2000, which would mean that the rate of growth of coal consumption will average 4 percent in the intervening period. Even to maintain its present share of the world energy supply will require a rate of growth of 3.5 per year, and some question exists as to how this can be done. On a worldwide basis there has been a marked decline in labor productivity in coal mining, and nowhere is this more obvious than in the United States, which is the second largest coal producer in the world (after Russia). Table 5-9 shows present production capabilities and gives some estimates of future capacities; but before examining it, the reader should be cognizant that although all types of coal are included in this table, the unit of measure is *hard* (or *black* or *anthracite*) coal, which is the most important component of total production and consumption and which dominates the international trade in coal. In terms of heating or calorific value, 1 ton of anthracite coal is equal to 1.0325 tons of *bituminous* coal.

It is also interesting to know that technically feasible levels of coal production are even higher than those indicated in these figures: the estimates are 4,500 Mtons in 1985 and 7,400 Mtons in the year 2000. Some marked shifts may also take place in the pattern of production. Canada could eventually become a very important exporter of coal, because although Canada's coal reserves are not very impressive, its total coal resources might turn out to be extremely large. It has also been suggested that China's coal resources are enormous, although Russia's are certainly larger. By the latest reckoning, the USSR had 48 percent of the world's coal resources and 17 percent of *reserves* (which, as was pointed out earlier, are *exploitable* resources). China has about 14 percent of the world's resources and 16 percent of

Table 5-9
Forecasts by the World Energy Congress and the International Energy Agency for the Production of Coal
(Mtons of hard-coal equivalent)

	World Energy Congress						International Energy Agency			
					Growth Rates				Growth Rates	
	1975	1985	2000	2020	1975-1985	1985-2000	1985	2000	1976-1985	1985-2000
Australia	69	150	300	400	6.5	2.9	109	285	5.1	6.6
Canada	23	35	115	200	4.3	8.2	40	71	8.0	3.9
China (PR)	349	725	1,200	1,800	7.6	3.4	—	—	—	—
Czechoslovakia	80	93	100	110	1.5	0.5	—	—	—	—
Germany (FR)	126	129	145	155	—	0.8	124	125	−0.2	0
Germany (DR)	75	80	90	100	0.7	0.7	—	—	—	—
India	73	135	235	500	6.3	3.8	—	—	—	—
Korea (North)	34	36	40	50	0.6	0.7	—	—	—	—
Poland	181	258	300	320	1.0	2.0	—	—	—	—
South Africa	69	119	233	300	5.6	4.6	—	—	—	—
United Kingdom	129	137	173	200	0.6	1.6	111	120	0.8	0.5
United States	581	842	1,340	2,400	3.8	3.1	837	1,181	4.7	2.3
USSR	614	851	1,100	1,800	3.3	1.7	—	—	—	—
Total	2,403	3,512	5,241	8,178	3.9	2.7	—	—	—	—
Other Countries	194	186	409	511	—	5.4	—	—	—	—
Total	2,597	3,698	5,650	8,689	3.6	2.9	—	—	—	—
OECD	—	1,394	2,185	—	5.7	3.4	1,329	1,928	—	—

reserves, while the United States has 25 percent of resources and 28 percent of reserves.

The important thing about the United States in this context is that its reserves are relatively easily won. The rail transport network in the United States is fairly complete, and there are no great physical or technical barriers in the path of its extension. This is not to say, however, that a sharp rise in coal production in the United States would not result in enormous pressures being placed on its transportation system. One of the reasons would be a movement of the center of gravity of U.S. coal mining to west of the Mississippi river, where by the year 2000 as much as 60 percent of U.S. coal mining might be taking place. At present this figure is 17 percent.

Most U.S. coal production now takes place in the East in underground operations. In the Western operations, open-pit mining or strip-mining dominates, which has meant either considerable environmental damage or an appreciable increase in the cost of the coal in order to cover restoration of the environment. Even so, costs are generally much lower for strip-mining than for underground mining; in particular, output per worker-day is usually much greater in strip-mining than in underground mines. In the United States, labor productivity is strip-mining may be as much as two or three times that of underground mining, and it appears that coal obtained from surface operations is priced at about one-half that taken from underground mines. In 1976 production in the United States came to about 600 million tons (Mtons), and on the strength of some scenarios developed by the Energy Modeling Forum at Stanford University, this figure could reach 2,600 Mtons in the year 2000. The determining factor here is seen as domestic demand since, according to the Stanford forum, neither reserves nor the transportation system in the United States provides an impediment. (However, certain financial hinders will have to be overcome; these are alluded to later.) About one-tenth of U.S. output is exported, with the major purchasers being Japan (which takes about 40 percent), Western Europe (about 20 percent), and Canada (about 25 percent). Obviously Canada imports U.S. coal because in many cases the coal is more easily shipped to Canadian buyers from the United States than from Canadian mines in a remote part of the same country, while Canadian coal from mines in the western part of Canada is more easily shipped across the Pacific to Japan.

In the Soviet Union, production increased during the 1970s by about 2 percent a year. This industry, however, has not been free of problems. In 1979 production was 3 Mtons lower than in 1977, amounting to 719 Mtons, and 33 Mtons below the target output. (Oil also experienced some setbacks: the total output in 1979 was 586 Mtons, which was an increase of 2 Mtons over the previous year but still 7 Mtons under the goal specified in the latest plan.) A clarification of these reversals is not available at present, but unofficial sources say that the Soviet transportation system is still inadequate, particularly if the winter is severe. Most Russian coal production takes place in the eastern part of the country, and more than one-half of Russia's export of coal goes to Japan.

Estimates of China's coal reserves are adjusted upward virtually every year; as yet, no one is willing to guess just what the ultimate figure will be. Chinese production reached 400 Mtons in the mid-1970s, and output is expected to grow steadily. The largest coal deposits are in the northern part of the country, but other regions also have sizable deposits. In recent years China has exported small amounts of coal to Japan, but since that country is intent upon diversifying the sources of its imports, this trade could grow. For it to reach major proportions, however, transport facilities from Chinese mines to ports would have to be greatly improved.

The International Trade in Coal

In heating value, the international trade in coal comes to only about 5% of the trade in oil and althogether comprises approximately 100 Mt/yr of coal. In 1978 prices, this coal has a value of about $8.6 billion most coal-producing countries are primarily concerned with domestic requirements, and very few are working to build up an overseas clientiele. Even two of the major coal-producing countries, Britain and West Germany, will probably not become major coal exporters in the foreseeable future. West Germany will probably continue as a minor exporter of coking coal to some of its steel-producing neighbors, but at the same time it may become an importer of steaming coal. In addition, although loading facilities are available in a few of the coal-rich countries for very large coal carriers (for example, South Africa and Australia), comparable facilities are conspicuously absent in the large coal-consuming countries. Japan seems to be the most prominent growth market for imported coal. The largest portion of Japan's coal imports is coking coal, mostly from the United States, Canada, and Australia; but is is reckoned that imports of steaming coal will increase from 0.4 Mtons in 1975 to 14 Mtons in 1985. The driving force behind this increase is uncertainty about the oil price.

Within the European Economic Community (EEC) as a whole, imports are expected to grow as a result of the general desire to reduce the dependence on oil; and in the shift of emphasis from oil- to coal-based power stations, considerably more steaming coal is going to be required than was thought necessary a few years ago. Present estimates are that total coal imports into the EEC in the near future are going to reach 85 Mtons, which is twice the 1976 level. It is interesting to note that the IEA has made some calculations of the cost of producing electricity in new oil, coal, and nuclear plants which could be put into operation by 1986. These calculations indicate that coal-using power stations are, from an economic point of view, clearly superior to oil-using ones. Table 5-10 summarizes this material and deals with power plants going into operation in 1986. The coal- and oil-burning plants are assumed to be equipped with filters or scrubbers designed to reduce atmospheric pollution.

In the United States a concentrated effort has been made over the past few years to get utilities to switch from oil to coal, but apparently to no avail. Now the President of the United States is about to propose that utility companies receive $12 billion over the next ten years if they reduce their use of oil by about 1/Mbbl/d by 1990. Although in the East this conversion is already taking place, many power companies are unenthusiastic about this program, arguing that it will cost too much. At the same time, local environmental agencies still feel that an increase in coal use will increase air pollution, since coal users might find some way to bypass the existing environmental regulations.

Most of the coal entering international trade is coking coal (for use in steel production). In the OECD countries, coke ovens consumer about 30 percent of hard coal, and other industrial used take 10 percent. Most of the rest goes to electricity generation, which also consumes 90 percent of the output of soft coal. The price of coal has risen steadily over the last decade or so, although the dampening of the international business cycle after the 1973-1974 oil price rises caused it to slow down somewhat. Coal is sold both on the spot market (mostly for immediate delivery) and via long-term contracts involving delivery over a period of ten to twenty years. In 1977 the price of coking coal on long-term contracts was between $50/ton and $60/ton, while its spot price was somewhat lower. This arrangement merely

Table 5-10
Estimated Average Cost per Kilowatthour of Producing Electricity in a New Power Station Using Different Types of Fuel, Assuming the Power Station Will Be in Operation 5,500 Hours per Year.
(*U.S. dollars, employing 1976 prices*)

	All Regions				
Cost	*Nuclear 2 x 1,100 MW*	*Oil (High Sulfur) 2 x 600 MW*	*Japan 2 x 600*[a]	*Western Europe 2 x 600*[a]	*United States 2 x 600*[b]
Average Capital Cost	14.9	9.6	12.9	12.4	12.4
Average Operation and Maintenance Cost	2.4	4.2	5.1	5.1	5.1
Average Fuel Cost	6.5	29.0	17.8	18.5	12.8
Total Cost	23.8	42.8	35.8	36.0	30.3

Source: International Energy Agency.
[a]Imported coal.
[b]Domestic coal.

reflected expectations about the future availability of coking coal, which in certain circles is considered inadequate in light of projected future demand. The price of steaming coal generally tends to average about one-half that of coking coal, varying considerably with respect to sulfur content and place of export.

As might be expected, transport costs are an important factor in determining the competitiveness of coal. Some of these are shown in table 5-11. Here it should be emphasized that the cost of sea transport could be decreased considerably if larger ships could be used. But, as pointed out above, the principal constraints on the use of these vessels are loading and unloading facilities at the major ports where coal is shipped and received.

A little known but interesting fact is that some of the highest-quality coal deposits in Europe are offshore, for the most part in the North Sea. The large gas deposits in the southern part of the North Sea are formed from coal. As yet, the technology for mining underseas deposits is inadequate; but a breakthrough in the *in situ* gasification of coal, which some engineers believe to be pending, could have important ramifications for Europe's energy supply. (Coal gasification is taken up in the last part of this chapter.)

Two more points need to be made before we leave the topic of coal. First, in order to expand the world coal production to 9 Gtons/yr by the year 2020, which is a goal often mentioned in light of decreasing oil supplies and in reality means only that coal's present share of the world energy market will be maintained, in the United States alone almost 400 new coal mines must be opened in the next twenty years; 160,000 new workers must be recruited to a not particularly popular profession; 5,000 new railway trains of various descriptions must be procured; large number of coal

Table 5-11
Some FOB Prices and Shipping Costs for Coal

	FOB Price ($/ton)	*1978 Ocean Transport Cost ($/ton)*	
		Europe	*Japan*
United States		4.75- 6.50	8.50-10.75
Hampton Roads/Norfolk	40.25		
Baltimore	33.50		
Poland		3.50- 4.00[a]	9.50-10.00
Gdansk	27.00		
South Africa		11.00-13.00[a]	9.00-11.00
Richards Bay	20.50		
India			
Haldia/Paradip	21.50		
Australia		12.75-14.00	6.50- 8.50
Newcastle/Port Kembla	30.50		

[a]To Italy.

barges and thousands of trucks must be obtained; at least nine pipelines need to be constructed; and investments amounting to at least $120 billion must be made over the first part of this period. Still, the United States will be producing only 25 percent of the world's coal.

Second, there is no point in talking about coal without talking about its environmental problems. To begin, almost all these problems can be solved, though at a cost which is very high on occasion. If we take strip-mining, land-reclamation techniques are now capable of satisfying the most fervant environmentalist. This has been shown at the Black Thunder development in Wyoming and at various sites in the Rhine area in Germany. Disposing of solid wastes from underground operations has been a more complicated matter, but now it appears that some of these solid wastes may have industrial uses: for example, fly ash, once considered an unmitigated nuisance, can be used to make cement.

The worst problem, of course, is pollution, particularly from sulfur oxides. It has been suggested, though, that the proportion of high- and medium-sulfur coal in total coal output will decrease, given the huge supplies of low-sulfur coal. As for the possibility of filtering out objectionable pollutants, it has been estimated that the installation of regenerable flue-gas desulfurization capacity in the United States would cost at least $35 billion (in 1978 dollars) if atmospheric sulfur concentrations are to be held to an acceptable level. However, so-called fluidized bed combustion eliminates a considerable portion of the most dangerous pollutants literally at the point of burning the coal. Obviously, though, development and capital costs for this new equipment are going to be tremendous, and unfortunately a great deal of effort will probably be made to avoid adopting it.

Natural Gas

In contrast to the growing uniformity exhibited in the estimation of oil reserves, there is a marked variability in estimates of the ultimate availability of natural gas. Proved reserves of natural gas are put at a minimum of 65×10^{12} m^3 (= 2,275 Tft^3), but recent estimates of ultimately recoverable resources of natural gas go up to 12,000 Tft^3, which is equivalent to more than 2 Tbbl of oil in heating value. In the matter of equivalences, it should be noted that one $m^3 = 35$ ft^3, and in heating value (or energy content) 1 ft^3 of natural gas is equal to 1,035 Btu. Thus one m^3 contains 36,225 Btu, and since the energy content of a barrel of oil [= 42 gallons (gal)] is 5,800,000 Btu, we see that 65×10^{12} m^3 is equivalent to $65 \times 10^{12} \times 36{,}225/5{,}800{,}000 = 405$ Gbbl of oil. (Thus, 1 barrel of oil = 5,600 cubic feet of gas.) A large part of the known reserves of natural gas are located far from present and potential markets; even more important, the means for receiving and transporting gas in the customer countries constitutes a significant barrier to more widespread use of this source of energy. Still, at present natural gas provides the world with about 17 percent of its energy supplies.

As compared with the other dominant fuels, natural gas is a late entry on the energy scene. The reason is that gas very often is found in the same geological structure as oil and to some extent can be labeled a by-product of oil. Thus, except in the United States and the USSR, natural gas was produced far from the potential centers of demand, as was oil; but unlike with oil, very high transportation and distribution costs per unit of energy content are encountered for gas: it costs five times as much to transport liquefied natural gas between the Persian Gulf and Japan as to transport oil with an equivalent energy content. Eventually, however, it became possible to link many areas where gas was being produced with energy markets by overland pipelines. This sort of thing occurred in Eastern Europe, Venezuela, and Mexico; and the development of technologies for liquefying and transporting refrigerated liquid natural gas has enabled the potential market for this commodity to be widened considerably. It should be emphasized, though, that these technologies involve more complicated and expensive facilities than are required for handling oil, and some environmental hazards are involved as well. At the same time, one of the principal attractions of gas is the absence of undesirable emissions into the atmosphere when it is burned, and consumers do not have to provide storage facilities.

In relation to production, there is no shortage of reserves of natural gas. As mentioned earlier, verified reserves are more than 2,000 trillion cubic feet, (Tft^3) and this is at least forty times the present yearly consumption of gas. However, U.S. reserves are falling rapidly and at present amount to about only ten times annual consumption. Western Europe has a reserve/production level of 22, but this may change somewhat as a result of the discovery of new reserves in the northern North Sea, particularly off Norway. A large gas find off Norway has recently been made by Shell. This brings Norwegian natural-gas reserves up to about 1,200 billion cubic meters (Gm^3), the largest in Western Europe. In my book *Scarcity, Energy, and Economic Progress* I suggested that it would be to the benefit of all concerned if the North Sea energy owners could become self-sufficient in energy as soon as possible and put themselves in position to supply Western Europe with at least a small amount of oil. But now it appears that North Sea gas may mean much more for the Western European energy picture than North Sea oil, although for reasons given later it may take some time for this gas to become available. We can now examine the reserve and production picture for natural gas as of 1977 (see table 5-12).

The United States is perhaps first among the countries in which natural gas has played an extremely important role in total energy supply. The energy content of the gas used in the United States is about the same as the energy content in U.S. oil production; and about half of the energy used in the household, trade, and industrial sectors has its origin in natural gas. The

Table 5-12
The Reserves of Natural Gas (1977) and the Production of Natural Gas (1976)

	Reserves		*Production*	
	Tft^3	*Percent*	*Gm^3*	*Percent*
USSR	918.0	43.4	336.0	20.6
Iran	330.0	15.6	50.0	3.1
United States	216.0	10.2	583.8	35.8
Algeria	125.8	5.9	21.5	1.3
Saudi Arabia	86.0	4.1	47.2	2.9
Holland	61.9	2.9	96.6	5.9
Canada	58.3	2.8	100.5	6.2
Nigeria	44.0	2.1	22.6	1.4
Venezuela	40.7	1.9	—	—
Australia	32.3	1.5	6.4	0.4
Kuwait	31.7	1.5	—	—
United Kingdom	30.0	1.4	39.4	2.4
Qatar	27.5	1.3	—	—
Iraq	27.0	1.3	—	—
Libya	25.8	1.2	—	—
China	25.0	1.2	17.0	1.0
Indonesia	24.0	1.1	7.5	0.5
Norway	18.5	0.9	4.4	0.3
Pakistan	15.8	0.7	—	—
Malaysia	15.0	0.7	—	—
Total	2,114.9	100.0	—	—

Source: *Basic Petroleum Data Book*, 1977; M. Folie and G. McColl (1978).

major reason is the comprehensive linking of the major centers of supply with the large urban and industrial centers by overland pipeline. The length of gas pipeline increased from 400,000 kilometers (km) in 1935 to more than 1,300,000 km in the late 1960s. The average diameter of pipelines has also increased, which means that more gas can be carried per mile of pipeline. According to Folie and McColl (1978), the accounting cost of the U.S. pipeline is approximately $17 billion.

But now it appears that reserves of natural gas in the United States have not increased in step with consumption. The reserve/consumption ratio has been steadily decreasing, and apparently production has peaked. It has been said that this latter phenomenon is the result of the regulation of the gas price, since the search for new gas reserves has slowed considerably. However, the demand for gas has been maintained because of its moderate price with respect to that of oil and because of environmental regulations, which have had the effect of increasing the relative cost of using oil and coal. It would appear that while the regulation of the price of U.S. gas has had a deleterious effect on its supply, there is no real validity in the claim that if this regulation were completely abolished, the natural-gas industry would materially increase its reserves in the short run.

Before 1973 it was claimed that the elasticity of supply of new gas reserves was as high as 2 (and thus a price rise of 25 percent would mean an increase in supply of 50 percent), but recent estimates are considerably less sanguine. As far as the DOE is concerned, the production of natural gas in the United States will continue to decline, and if more gas is desired, imports will have to increase. In all fairness, though, it must be recorded that higher gas prices have led to a great deal of new drilling, and although production is low, so are costs, while the discovery rate is high. Similarly, decontrol has meant increased geological interest in several regions which ostensibly have high potential. The most attractive is the Western "overthrust" belt where, in a strip 50 miles wide running from the Canadian to the Mexican borders, some geologists say as much as 200 Tft^3 of gas may be located. In 1975 the United States imported 1.35 Tft^3 from Canada, which was 40 percent of that country's production; but at present Canadian production has started to decline, and the United States is seeking long-term agreements on gas deliveries from a number of other countries.

The Soviet Union occupies a special place on the world gas scene. Since 1974 they have been a net exporter of gas, although earlier they imported sizable amounts from Iran and Afghanistan, and they may do so again. Their exports in 1976 came to 26 Gft^3, divided evenly between Eastern and Western Europe and transported by pipeline. It has been said that these exports may double during the coming decade when the rich Siberian fields are exploited, and among others both Japan and the western United States are spoken of as being potential buyers of Siberian gas.

A major problem for Russia regarding the supply of natural gas is its enormous area. Thus the 98,000 km of pipeline owned by that country is not enough to prevent some important regional supply imbalances. There has been some contention that the expansion of this pipeline has consistently fallen below planned levels, but in 1979 the gas industry seems to have performed much more satisfactorily than the other energy industries, and output may have grown by as much as 9 percent. Altogether, natural gas supplies 25 percent of Soviet energy consumption, with 10.5 percent of this being used for household purposes, 51.1 percent consumed in industrial activities, and 30.8 percent used in the generation of electricity. The principal difficulty faced by the Russian energy sector is the shift in its center of gravity toward Siberia, where enormous engineering feats are going to be necessary before the resources of that region can be efficiently exploited.

The gas industry of Western Europe originally was based on gas manufactured from coal, but gas was discovered in the Po Valley in Italy, in 1940 and sometime later around Lacq in France. The largest discovery, however, was Holland's Groningen field. It has been suggested that without these supplies Holland might have found itself in desperate straits in the

1970s, given the absurd extent to which the Dutch have taken some of their welfare programs: 51 percent of Holland's energy supply is based on gas, and that country is also a large exporter of gas, via pipeline, to surrounding countries. But in the mid-1980s these exports should start to decrease very rapidly, and by 1995 will be very modest. As was mentioned earlier, it is possible that gas from Norway and Britain can replace some of the declining Dutch supply, but a problem here is that gas in the northern part of the North Sea is, for the most part, a by-product of oil and thus a function of the amount of oil produced. The Shell find mentioned earlier may be in this category, although there is some evidence that the oil associated with this gas is "dead oil" which may not be profitable to exploit. Only a few countries possess large concentrations of gas without oil, with Canada, Australia, Western Europe (excluding the northern North Sea), Japan, and Chile being the most notable. Even Algeria, which at one time was in this position, now finds that new gas reserves are associated with oil.

We conclude this part of the discussion by saying something about Japan and OPEC. The existing high levels of atmospheric pollution have turned Japan into a good market for natural gas, especially in the major population centers. In 1977, 3.5 percent of Japanese energy consumption was in the form of gas; but because of the badly developed gas-distribution system, it seems unlikely that consumption can grow by a large amount. The powerful Ministry of International Trade and Industry at one time prepared a plan calling for increasing the imports of gas into Japan, with the supplying areas being (first and foremost) Australia, Indonesia, Brunei, Malaya, and some of the Middle Eastern countries. But this plan seems to have been shelved, at least for the time being. For the Japanese, the principal stumbling block seems to have been the necessity to invest heavily in an expensive distribution system without knowing exactly either how much gas will be available in the future or the price of future supplies. In particular, the Japanese authorities are reluctant to build an expensive distribution system to carry foreign gas to households and small industries, instead preferring to direct their imports of this commodity to large-scale energy users.

As for OPEC, many of these countries have simply *flared* their gas in the past (which means that they set it afire, thereby completely destroying its energy content), but at present there is increasing interest in using it in one sense or another. Algeria, Iran, Libya, Nigeria, Abu Dhabi, and Indonesia have expressed a strong interest in increasing exports, while some of the others (such as Saudi Arabia) also have indicated a desire to increase local consumption. It would be quite logical for Saudi Arabia to begin thinking in terms of developing a petrochemical industry which could use local natural gas as a feedstock. This is particularly true now that the supply of Saudi engineers and technicians is increasing and there are no financial constraints on the establishment of such an industry, and the government of that country is anxious to develop a more versatile economic base.

International Trade

The cost of transporting natural gas is one of the main barriers to an expanded trade in this commodity, and this is true whether we are talking about sea transport or transport by pipeline. If we take transport by sea in the form of liquefied natural gas, then the delivery of about 10 Gm^3/yr by Algeria to the United States would require an investment of $2.6 billion, broken down as follows: $1 billion for a liquefaction plant to convert the gas into a cold liquid, $1.3 billion for eight cryogenic tanker ships, and $300 million for a regasification facility in the United States. A rough estimate of costs often used is that a liquefied natural gas (LNG) system capable of supporting an annual trade in gas equivalent to 6 to 8 Mt/yr of oil would require about $3 billion.

The capital requirements for LNG projects are so large that in arranging their financing it has become usual to think in terms of twenty-year agreements with indexed prices. The capital cost per unit of energy for an LNG project involving almost any of the present nonarctic operations exceeds that of North Sea oil, which is commonly regarded as some of the most expensive oil in the world.

In trying to get an estimate of future world trade, we can start with some predictions of domestic demands and supplies. If we take the case of the United States, we see little possibility of obtaining large new supplies in the lower forty-eight states. The only potential source of large new gas supplies is Alaska, and even with pipelines from that region to the lower forty-eight states there is going to be a shortfall of gas which can be met only by imports. A large part of these imports may arrive by pipeline from Canada, and the rest come from sources mentioned earlier in this section.

A considerable excess demand also seems scheduled for Western Europe. Exact estimates vary considerably, although it has been suggested that by the year 2000 massive imports into Western Europe may be necessary. A large part of these imports could be provided by pipeline from the USSR, assuming that the Soviet pipeline system could be enlarged and the world political situation does not deteriorate; OPEC undoubtedly has the potential to supply the rest and desires a financial return on the gas now being flared. Some question has been raised as to whether OPEC would attempt some kind of cartel-like behavior with gas, but generally this possibility has been rejected. Gas is not as important as oil, and the high transport and capital costs associated with gas, which vary according to points of export and import, should prohibit a uniform FOB price among potential cartel members. However, Saudi Arabia may soon argue that given the future market for gas, they cannot continue to flare this product. This might lead to a decrease in their present output of oil, since in that country oil and gas are joint products.

Some Other Energy Resources and a Conclusion

There are other energy sources than those mentioned previously in this chapter, and some of these are much more desirable from both a political and an environmental point of view. The principal source is solar energy, which may someday remove a large proportion of our energy worries—but not yet. Although in June 1979 President Carter announced that by the year 2000 he believed that 20 percent of the energy used by the United States could be derived from renewable resources (solar, biomass, geothermal, water, and wind), he did not say how this was going to come about. In fact, his contention that some of the technologies associated with these energy resources verge on the economical might be misleading. Solar space heating can be up to five times as expensive as a conventional system, and solar cooling can be even more expensive. Large-scale applications of photovoltaic technologies (for transforming the sun's rays directly into electricity) are not satisfactory. Exxon calculates the cost of this kind of power in the United States at $1.3/kWh, while conventional power costs $.05/kWh to $.10/kWh. There is also the size of solar facilities. For a country like Sweden, replacing a medium-sized power plant having 6 TWh of electric energy by solar facilities would require a reflecting area of about 100 Km.[2]

It should be made clear, however, that it is probably true today that much more soft energy *should* be used than is actually being used, particularly at the household level. The major problem is the absence of comprehensive schemes for financing the introduction of these energy forms, particularly now, when household incomes over a large part of the industrial world are no longer expanding at the previous rate. Of course, if or when economical equipment which can store energy gathered from the sun becomes available, this problem may disappear, because then the gains from employing more solar energy will be obvious to even the most obtuse Gothenburg energy expert. This observation also holds for wind power. For Sweden, replacing a power station supplying 6 TWh would require 600 wind installations, each 100 meters (m) high and having a blade diameter of 100 m. However, in Sweden, the northern part of the United States, Canada, and so on, small wind power plants for the household sector should eventually prove to be extremely useful, particularly when satisfactory devices for storing the generated electricity are available. At present there is a widespread belief that such storage devices (as well as economical photovoltaic technology) will be available by the turn of the century at the latest. But even in a country such as Sweden, where there is widespread support for so-called soft-energy options, nobody has bothered to promote the financial incentives that would encourage individuals and firms to accelerate the development of the equipment required for this option.

Before we leave this topic, it should be pointed out that there have been many promising experiments with wind power. From 1958 to 1962, an 800-kW wind-powered plant was connected to the electricity supply system at Nogent-le-Roi (near Paris) and was dismantled only when the price of oil fell to and remained at an unusually low level. In Denmark a 2,000-kW plant (the world's largest) has been put into service at Ulfborg, and apparently it is supplying power at a cost competitive with those of other energy sources. Similarly, the Netherlands is studying the possibility of supplying up to 20 percent of Holland's electric power by windmills. In the United States both the DOE and many utility companies feel that wind will soon be an important energy source.

In the business world in the United States, where it would appear that the gravity of the oil problem has at last overwhelmed some of the more absurd energy economics offered by various celebrities within the academic world, the great hope is synthetic fuels (or, to be more correct, synthetic oil). First and foremost, we are talking about turning the huge supplies of coal in the world into oil. For example, 1 ton of U.S. coal can produce 2.5 bbl of oil, and in South Africa 20,000 bbl/d of oil is now being produced from coal at a cost of about $25/bbl. (However, South African coal costs much less than U.S. coal.) It also seems to be the case that 1 ton of South African coal produces between 1 and 2 bbl of oil.

Three main processes are relevant here. The first involves reducing the coal to a fine powder and, after a chemical solvent is added, processing the resultant mixture to increase its hydrogen content. In this state it can be converted to a liquid. The second process is based on burning the coal to produce a gas and then liquefying the gas. In the United States, commercial processes already exist for underground burning of thick seams of coal located fairly close to the surface. This underground, or *in situ*, gasification requires as much capital as underground mining and generates a gas that is poor in heat content. But it does avoid a great deal of atmospheric pollution. This is very important, for in a surface operation that converts coal to oil, a large amount of the coal disappears into the atmosphere in the form of carbon dioxide and other noxious pollutants. Third, there is an important process for converting coal directly to methyl alcohol. In the South African process, coal, steam, and oxygen are combined under high pressure and temperature to yield a gaseous mixture of carbon monoxide, hydrogen, and methane. Then the gasses, under pressure and at high temperatures, are passed over an iron-based catalyst to produce oil and fuel gas. The oil can be refined into gasoline. One of the more interesting aspects of the South African process is that the oil is said to be of a better quality than conventional crude oil because of its extremely low sulfur content.

In addition to his solar ambitions, President Carter wants 2.5 Mbbl/d of synthetic oil by 1990. (The United States is now using almost 20 Mbbl/d of oil.) It is said that facilities for producing 2.5 Mbbl/d will cost a minimum of $45 billion, with the oil selling for at least $35/bbl. (Pilot plants operated by Gulf Oil and Exxon now produce 30,000 bbl/d of oil for a cost of approximately $30/bbl.) A huge increase in coal production would also be required in order to supply feedstocks for these facilities, and thus the actual price of this kind of oil would probably exceed $50/bbl if a program approaching the desired output were realized. Even so, many energy executives in the United States feel that an accelerated synthetic-fuels program is a must and will be initiated sooner or later. Among other things, it is pointed out that had 1973-1974 intentions for synthetic oil come to fruition, it would have been possible for the synthetic-oil industry to produce one-half of the U.S. shortfall of oil at the time of the Iranian revolution.

They also feel that if a program is not undertaken, and sustained, on a fairly large scale, then world energy prices are placed squarely in the hands of OPEC, and apparently there is a possibility that their interests may not coincide with those of the industrial world. What we have here, in fact, is a variation of a game known as the *prisoners' dilemma* in which great rewards are avilable to the players if they cooperate, but they are also enormous penalties for anyone making the mistake of trying to be nice if the other player decides to be nasty. The big problem is that when the industrial world thinks of cooperation, it is thinking in terms of maintaining its extremely high and, to a certain extent, unnecessary consumption of energy; on the other hand, OPEC probably thinks that it is performing a service to humanity by keeping prices high and thus "encouraging" its clients to switch to new energy sources and to accelerate conservation. The outcome of this palpably different way of viewing the same phenomenon guarantees both economic and, eventually, political turmoil throughout the entire world, over the indefinite future. A possible compromise would be for those OPEC countries with spare cash to help finance energy projects in the industrial world in return for either equity positions in these projects or a slice of the flow of goods and services that these projects would facilitate. For this arrangement to work, though, there would have to be a measure of confidence between the two sides that is apparently lacking today and, thanks to certain actions taken by the U.S. government in the wake of the Iranian revolution (which involved the freezing of some Iranian assets in the United States), can hardly be expected to surface in the near future.

Heavy Oil, Tar Sands, and Shale

The expression *synthetic oil* is often taken to mean oil produced not only from coal, but also from tar sands and shale; and some investigators also consider oil from tar sands a heavy oil. This matter is considered a little more closely in this section, but it can be noted at once that oil shale does *not* contain oil and, strangely enough, is not even a shale. (Oil is produced by destructive distillation of the organic matter in this material, and this organic matter is called *kerogen*.)

As alluded to earlier in this book, the American Petroleum Institute (API) maintains an index of the quality of crude oils based on a simple viscosity test. One Algerian light oil has an API rating of 44°, while tar-sand oil has an API rating of 8° to 10°. On the other hand, one Venezuelan heavy oil (Bachequero) has an API of 16.8°. Thus we might suspect that both heavy oil and oil from tar sands have various characteristics which distinguish them from conventional oil, and this suspicion, if it exists, is justified. Heavy oil, for example, was once rejected as a source of energy, but the oil price rises that began in 1973 resulted in a reappraisal of its usefulness.

Although some heavy oil is found in shallow reservoirs, it has the consistency of molasses, as well as a high sulfur content and less potential energy than the lighter grades of oil. Moreover, because it is so viscous, less than 10 percent of a typical deposit will flow to the surface if it is the object of conventional pumping methods. In general, it must be liquefied by being exposed to superheated steam for as much as two weeks, at pressures up to 2,500 pounds per square inch (lb/in^2). This, by itself, requires a great deal of energy: to get 3 bbl of oil can require an energy equivalent of up to 1 bbl. Refining heavy oil is also a complicated business. Heavy-oil molecules have a higher proportion of carbon atoms, and fewer hydrogen atoms, than conventional light crudes; and since the energy potential of oil depends on its hydrogen content, a large proportion of refined heavy oil takes the form of products with a low energy potential, such as industrial fuel oil and bunker oil for ships. Obtaining more gasoline from a refinery slate requires a larger-than-normal investment in processing equipment.

The principal bonus of heavy oil is quantitative. Venezuelan reserves of heavy oil may be as large as *all* the reserves of light oils in the world; and the United States may have enough heavy oil to double its total reserves. It should also be made clear that for purposes of classification, any oil with an API rating of less than 20° that "flows" into production lines can be considered a heavy oil. By this criterion tar sands are *not* a heavy oil, since under no circumstances do they flow unless heated or diluted.

Tar sands consist of sand grains loosely bound together by a petroleum substance called bitumen; like heavy oil, bitumen is light on hydrogen and heavy on carbon (83 percent by weight). These sands occur mainly in Albania, Rumania, the USSR, Colombia, Venezuela, and the United States; but the major world reserves (perhaps 710 Gbbl of oil in place) are to be found in Canada. With the present recovery rate estimated at between 30 and 50 percent, the Canadian tar sands could eventually yield at least 200 Gbbl of oil, although present production is quite modest (about one-fourteenth of Canada's oil requirements). Some oil is also being produced from tar sands in Albania, Rumania, and the Soviet Union; and it has been said that both Colombia and Siberia may have deposits of this resource equal to or greater than the main producing area of Canada—the Athabasca tar sands district of northern Albera, which contains the largest known petroleum deposit on earth.

One of the more interesting aspects of tar sands is that they are the poorest quality petroleum resource being developed anywhere in the world; but because of the skyrocketing price of conventional oil, they have become extremely valuable. The problem here is that oil literally has to be "mined" from tar sands, as described in Banks (1976). The development of the Athabasca deposit has been complicated and slow, and operating costs have escalated in an unforeseen manner. It has also been said that the direct and indirect energy input needed to produce a barrel of synthetic oil from tar sands is even larger than that cited above for heavy oil. By contrast, the energy gain from conventional oil is on the order of 100 to 1. It also seems that few oil-producing operations anywhere in the world are more expensive than that of Syncrude Canada Ltd. Capital investment for each daily barrel of production capacity is $20,000, while this cost in the Middle East is only $350. Estimated in 1978 dollars, production cost per barrel will be $9.50 for Syncrude's output, while comparable production costs for conventional oil are $4.50 in the United States and from $.25 to $1.25 in the Middle East. Still, it would be difficult to claim that at the present and expected price of oil (and present and expected escalations in operating costs), the Syncrude project is not profitable. And some geologists claim that, because of the enormous size of the tar-sands deposits in Canada and the length of time over which they can be exploited at a high rate of output, the per-barrel cost may eventually be comparable with that of North Sea oil.

Finally, we can look at oil from shale. At present this commodity is produced in only a few countries, principally the Soviet Union and mainland China, and estimates are that about 35 to 40 M tons of oil shale is being mined every year. In the matter of economical reserves (or reserves from shale that contains more than 42 liters of oil per ton of mined material), the United States and Brazil occupy a special position. The United States has about 80 Gbbl of oil locked up in shale that falls in the category of reasonably ensured reserves, and perhaps as much as another 950 Mbbl could eventually be made available. Brazil has about half this amount; but there are also appreciable supplies in the Soviet Union, Central Africa, Canada, Sicily, and mainland China. In both China and Russia, for example, several shale-oil power stations are in operation. All told, "reasonably ensured" world reserves of shale oil amount to about 135 to 150 Gbbl, while ultimate reserves might come to 2,500 Gbbl.

The first thing to note about shale oil is the difference in quality of various deposits. This quality is usually defined by the oil content in liters per ton, in conjunction with the calorific value, where these are "approximately" correlated. Of the major shale reserves, those of Estonia (that is, the Soviet Union) have the highest quality. The shales of Brazil have about one-half the energy content of these Soviet resources; and Colorado shales have slightly more than one-third. Still, the total known reserves in the shales of Colorado, Utah, and Wyoming are one-half as large as the reserves of conventional oil possessed by Saudi Arabia, and total resources are probably much larger. Three tons of shale yields about 2 bbl of oil at a cost of $20/bbl to $40/bbl. (Gulf-Standard of Indiana has an operation planned in Colorado that will yield 200,000 bbl/d at $25/bbl. One pilot plant is already functioning.)

Shale rock, which in reality is hardened clay, has to be crushed and then heated to about 900°F in the process of obtaining oil. The U.S. government wants 500,000 bbl/d of this commodity from Colorado shale. Estimates are that this could involve disposing of several hundred thousand tons of solid waste a year because crushing the rock causes it to expand by more than 20 percent and thus it produces more waste than the hole which it has left could accommodate. Thus 500,000 bbl of oil necessitates the crushing of approximately 750,000 tons of rock; but this could involve in excess of 900,000 tons of waste, of which less than 750,000 tons could be pressed back into the original excavation. Shale also contains cancer-causing pollutants, and obtaining 500,000 bbl/d of oil would require about 5 percent of Colorado's river water. There is a possibility that some shale could be mined *in situ,* but as yet nobody knows enough about this technology to say whether it is economical or just what its environmental effects will be in comparison to those of surface mining.

If the disposal problem can be overcome, shale oil almost certainly is capable of providing an important increment to the world's energy supply. Although it uses some of the worst-quality shale in the world, the economics of the Portlandzementwerk Dotternhausen Rudolf Rohrback KG, in Germany, is impressive. At this plant oil shale functions both as a source of energy and, through the processing of oil-shale combustion residues, a raw material for cement production. In addition, processes are being developed for extracting valuable metallic resources such as uranium and vanadium from power station residues. The *lowest* estimate by British Petroleum for the recoverable reserves of heavy oils, and tar sands and shale oils, is 600 Gbbl, which is almost equal to the total of verified reserves of conventional oil.

A process that can also be mentioned here is petroleum mining. This is an underground process that, in theory, can free petroleum that is unrecoverable by other techniques, to include enhanced recovery. A spokesman for the U.S. Bureau of Mines claims that this method could potentially recover 40-50 percent of the oil remaining in a depleted field; and impressive petroleum mining operations are being carried out in Russia.

Appendix 5A: Constant-Elasticity Demand Curves and Dynamic Sequences

Earlier in this chapter there was a nontechnical discussion of the effect of price rises which employed constant-elasticity demand curves. Some of that discussion is generalized here, in a situation where we have a price fall. In both cases the concept to be grasped is that we have a larger long-run than short-run elasticity, where by "larger" we mean larger in absolute value. Using some figures presented earlier, we might go from -0.1 in the short run to -0.3 in the medium to long run; but in absolute values this is 0.1 to 0.3. The equation of our demand curve is taken as $D_t = \alpha P_t^{\beta}$, where β, the elasticity, is less than zero for a "normal" demand curve and P_t is the price. In logarithmic form, this is $\log D_t = \log \alpha + \beta \log P_t$. We see right away that β is an elasticity by differentiating:

$$\frac{dD_t}{D_t} = \beta \frac{dP_t}{P_t}$$

Thus

$$\frac{dD_t/D_t}{dP_t/P_t} = \beta \qquad \beta < 0$$

Now we need an adjustment relation. *With D^* as equilibrium demand,* this can be taken as $\log D_t - \log D_{t-1} = \lambda (\log D^* - \log D_{t-1})$, with $0 < \lambda < 1$. With the demand equation $\log D^* = \log \alpha + \beta \log P_t$, we have $\log D_t = \log D_{t-1} + \lambda (\log \alpha + \beta \log P_t - \log D_{t-1})$, or

$$\log D_t = \lambda \log \alpha + (1 - \lambda) \log D_{t-1} + \lambda\beta \log P_t$$

which is a well-known construction in econometrics. Note that we can get the short-run elasticity by differentiating this expression with respect to P_t, which yields $\lambda\beta$. To get the long-run elasticity first we set $\log D_t = \log D_{t-1} = \cdots = D^*$. Thus $\log D^* = \lambda \log \alpha + (1 - \lambda) \log D^* + \lambda\beta \log P_t$, or, by simplifying, $D^* = \log \alpha + \beta \log P_t$ (as before). Thus the long-run elasticity is β; and with $0 < \lambda < 1$, the (absolute value of the) long-run elasticity is larger than the short-run elasticity. However, we must still show that these manipulations correspond to the arrangement show in figure 5A-1, where we begin our story with a fall in price from P_a to P_b.

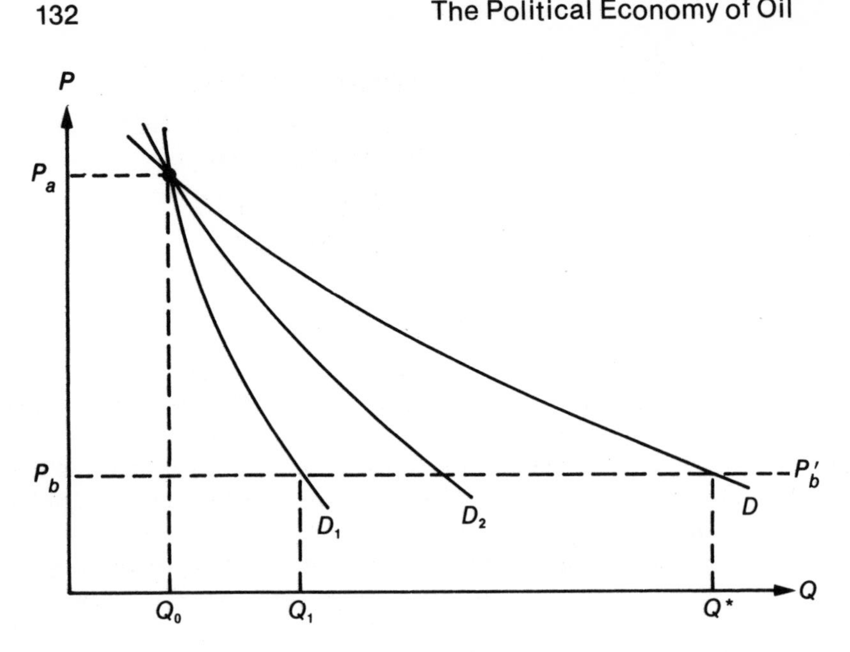

Figure 5A-1. Dynamic Demand Curves Having a Constant Elasticity

On a demand curve such as D_1 we must satisfy $Q_0 = \alpha P_a{}^{-\beta}$, where now, as compared to earlier, β is taken to be greater than zero. By the same token, we have $Q_1 = \alpha P_b{}^{-\beta}$. Thus we get

$$\frac{Q_0}{Q_1} = \left(\frac{P_b}{P_a}\right)^{\beta}$$

or

$$\log\left(\frac{Q_0}{Q_1}\right) = \beta \log\left(\frac{P_b}{P_a}\right)$$

Thus

$$\beta = \frac{\log Q_0 - \log Q_1}{\log P_b - \log P_a}$$

and

$$\frac{d\beta}{dQ_1} = \frac{-1}{(\log P_b - \log P_a)Q_1}$$

Since $\log P_b - \log P_a < 0$, when Q_1 increases, the *absolute value* of the elasticity β must be larger. Note here the meaning of the expression Q_1 *increases*: An increase in Q_1 means that it is farther to the right in the diagram. For instance, the demand curve D_1 would be in the same position as D_2, and thus the absolute value of the elasticity on this curve would be larger. Thus as we move to the right on the line $P_b - P_b'$, we are crossing demand curves whose elasticities are higher than those of the preceding curves.

To summarize, when the price falls from P_a to P_b (and remains at this level), demand increases from Q_0 to Q_1. This is an *impact effect*. As time goes by, demand begins to approach Q^*. Conceptually, we are moving from the short-run demand curve D_1 toward the long-run curve D, crossing some intermediate demand curves (such as D_2) as we do so.

We need to discuss dynamic sequences here, since the demand curves shown in figure 5A-1 are sometimes called dynamic demand curves. The basic issue is the movement from one equilibrium to another. In this figure we have an equilibrium at (P_a, Q_0) which is "disturbed" by a price change to P_b. This sets us in motion toward an equilibrium at Q^*, assuming that such an equilibrium exists; but all this takes time. If we had demand curves for ordinary household goods, force of habit would prevent a large price fall from being immediately translated into a large increase in demand. Similarly, a large fall in the price of fuel would not immediately lead to a large increase in the amount of driving, and thus a large increase in the demand for fuel.

Some of these observations can be clarified with the help of a little algebra using figure 5A-1 as a reference. If we begin with price P_a and demand Q_{t-1}, a fall in price to P_b will eventually result in a long-run, or equilibrium, demand of Q^*, But this position will be approached only gradually. In the first period, after the change in price, the buyers' intention is to increase demand by the amount $\lambda (Q^* - Q_{t-1})$. For instance, if $\lambda = 1/4$, $Q^* = 200$, and $Q_{t-1} = 100$, the increase in demand would be 25 during the initial period, bringing the total demand to 125. During the next period, the increase in demand would be $\lambda (Q^* - Q_t)$, and this would come to $1/4(200 - 125) = 18.75$. Total demand is thus $100 + 25 + 18.75 = 143.75$. In order to get total demand after the third period, the same procedure can be repeated, but at this point we formalize our exposition. Thus we have

$$\Delta Q_t = Q_t - Q_{t-1} = \lambda (Q^* - Q_{t-1})$$
$$\Delta Q_{t+1} = Q_{t+1} - Q_t = \lambda (Q^* - Q_t)$$
$$\Delta Q_{t+2} = Q_{t+2} - Q_{t+1} = \lambda (Q^* - Q_{t+})$$

and so on. Confining ourselves to two periods, we can write

$$Q_{t+2} = Q_{t+1} + \lambda [Q^* - Q_{t+l}]$$

$$= Q_t + \lambda (Q^* - Q_t) + \lambda [Q^* - Q_t - \lambda (Q^* - Q_t)]$$
$$= Q_{t-1} + \lambda (Q^* - Q_{t-1}) + \lambda [Q^* - Q_{t-1} - \lambda (Q^* - Q_{t-1})] + \lambda \ \{[Q^* - Q_{t-1} - \lambda (Q^* - Q_{t-1}) - \lambda [Q^* - Q_{t-1} - \lambda (Q^* - Q_{t-1})]\}$$

This manipulation involves continuous substitution from the previous set of relations in order to get everything in terms of Q^* and Q_{t-1}. Now, the last expression can be simplified to

$$Q_{t+2} = Q_{t-1} + \lambda (Q^* - Q_{t-1}) + \lambda (1 - \lambda) (Q^* - Q_{t-1}) + \lambda (1 - \lambda^2) (Q^* - Q_{t-1})$$

If we were to use the above numerical example, we would have

$$Q_{t+2} = 100 + \tfrac{1}{4} (200 - 100) + \tfrac{1}{4} \ \tfrac{3}{4} \ (100) + \tfrac{1}{4} \ \tfrac{3}{4} \ \tfrac{3}{4} \ (100)$$
$$= 100 + 25 + 18.75 + 14.0625 = 157.8$$

In other words, with the new equilibrium demand at 200, at the end of *three* periods (since we began at $t - 1$), demand has increased to 157.8. The proof that it will eventually reach 200 is simple, although reaching this level takes a great deal of time—infinity to be exact, as might be inferred from the development shown in figure 5-4. This proof involves simply extending and rearranging a series of the type shown above. Thus we get

$$Q_{t+n} = Q_{t-1} + \lambda (Q^* - Q_{t-1}) + \lambda (1 - \lambda) (Q^* - Q_{t-1}) + \cdots + \lambda (1 - \lambda)^i (Q^* - Q_{t-1}) + \cdots$$
$$= Q_{t-1} + \lambda (Q^* - Q_{t-1}) [1 + (1 - \lambda) + (1 - \lambda)^2 + \cdots$$

The series in parentheses can be summed in the usual way. As $n \rightarrow \alpha$, we get

$$Q_{t+n} = Q_{t-1} + \lambda (Q^* - Q_{t-1}) \frac{1}{1 - (1 - \lambda)} = Q_{t-1} + Q^* - Q_{t-1} = Q^* \qquad \text{Q.E.D}$$

The basic issue that we are dealing with here is distributed lags. If demand D depends on some lagged variable y, it is possible to write

$$D = \int_0^\infty \alpha (\tau) y(t - \tau) \, d\tau \qquad \text{with} \qquad 1 = \int_0^\infty \alpha (\tau) \, d\tau$$

As specified here, the weights sum to unity. Next, if these weights $\alpha(\tau)$ can be taken as exponential, or $\alpha(\tau) = ae^{-\lambda\tau}$, we have

$$1 = \int_0^\infty ae^{-\lambda\tau}\, d\tau = \frac{a}{\lambda}$$

Thus we get $a = \lambda$. Now we can write

$$D = \int_0^\infty \lambda e^{-\lambda\tau} y(t - \tau)\, d\tau$$

Taking $x = t - \tau$ gives $dx = -d\tau$; then, substituting in the above and rewriting, we get

$$De^{\lambda t} = \int_{-\infty}^t e^{\lambda x} y(x)\, dx$$

This can be differentiated with respect to time, to obtain

$$\dot{D} = \frac{dD}{dt} = \lambda(y - D)$$

We therefore see that the rate of change over time of D is proportional to the difference between the desired value of D, which is y, and the actual value of D. This is clearly the continuous analogue to the expression with which we started the previous exposition; and its diagrammatical counterpart can be seen in figure 5-4.

Finally note that we could have an entirely different demand structure, for example:

$$D_t = f(P_t, Pt^{e} + 1) = a_0 + a_1 Pt + A_2 Pe_{t+1} + _1)$$

with $a_1 < 0$ and $a_2 >$. The specification for a_2 means that an *expected* increase in the price of a product in the coming period will lead to an increase in present demand. As for an expectational scheme, let us take $Pe_{t+1} = P_t + \phi(P_t - P_{t-1})$. Thus if $\phi > 0$ and we have an increase in price from period $t-1$ to period t, a higher price will be expected in the following period. Using Pe_{t+1} from the previous equation gives us:

$$D_t = a_0 + [a_1 + a_2(1 + \phi)]\, P_t - a_2 P_{t-1}$$

Given the signs of a_1 and a_2, it is apparent that $a_1 + a_2(1 + \phi) >< 0$, depending on the magnitudes of a_1, a_2 and 0. Under the circumstances we could get a postive elasticity of demand.

Oil and Macroeconomics

From 1951 to 1973 the industrial world experienced a golden age composed of steady, unremitting social and economic progress. The average rate of economic growth for the OECD was 4.8 percent per year in real terms (versus the 3 to 3.5 percent or less that is likely to be the average for the immediate future). The rate of inflation was low or moderate in many industrial countries, which is a matter of no small importance for pensioners and others on fixed incomes, and long spells of unemployment were an experience generally reserved for people who chose it.

This largesse was due to a high degree of political stability; accelerating technical progress, based to a considerable extent on scientific techniques introduced or perfected during World War II; and, as we recently found out, falling real prices for energy, for nonfuel minerals, and for agricultural products. The real price of oil fell by one-half over this period, and the most important nonfuel minerals—copper, aluminum (bauxite), and iron ore—suffered similar declines in their terms of trade vis-à-vis industrial products. Of course, there were occasional downturns or recessions; but if these were not manmade in the sense that they were avoidable, at least they were "man-maintained," since in most countries both the government and the politicians had sufficient knowledge (at least in theory) about monetary, fiscal, and labor-market policy to mitigate the worst effects of these disturbances. The reader must be very careful to notice here that these disturbances were not based on a shortage of energy and natural resources. This phenomenon is quite new, and its alleviation is going to require a drastic departure from previous orthodoxy.

The international financial system also functioned with remarkable smoothness. The reader desiring a brief but comprehensive survey of this mechanism will find it in my book *The International Economy: A Modern Approach.* A key component was the system of fixed exchange rates, which reduced uncertainty in international trade and compelled governments to make some effort to keep external accounts in balance. The matter of uncertainty is probably clear to anyone who has taken, or contemplated taking, a holiday abroad. One of the first factors that must be settled prior to such a trip is the quantity of foreign goods and services that can be purchased with the currency of one's own country. This is determined by not only prices in the foreign country, but also the rate at which domestic currency exchanges for foreign currency. Foreign travel is somewhat less complicated when the traveler knows that if $1 is equal to 2 marks when the trip is being planned, then $1 will not equal 1.75 marks when the traveler arrives in Munich. Businesses face a similar dilemma. Scandinavian shipbuilders who, in the late 1960s, signed contracts specifying their sales prices in dollars were somewhat annoyed when, on delivering these ships in the early 1970s, they were paid in dollars which had depreciated by 10 to 15 percent in value (measured in local currency) because of the abandonment of the system of *fixed* exchange rate and the introduction of a system of *flexible* or *floating* rates. (We actually have a system of *managed* floating in which movements in the exchange rate are permitted as long as they are not extreme. This arrangement is often called "dirty" floating.)

For our purposes, an even more important example is the seller of oil who receives a certain number of dollars for a given amount of oil, but who finds that because of the fall in the value of the dollar with respect to the yen or schilling, the dollar will buy less Japanese or Austrian machinery than originally contemplated. This particular situation will undoubtedly have a considerable influence on the price of oil in the future (and perhaps the near future) since, for better or worse, several important OPEC personalities have indicated that they no longer find it tolerable, and they intend to take steps to see that the *real* purchasing power of their oil revenues is maintained.

The matter of the foreign balance requires a brief explanation. Under a system of fixed exchange rates, if a country persistently buys more goods and services abroad than it sells and cannot attract additional foreign exchange (that is, foreign money) by selling bonds and stocks to foreigners or by getting them to purchase local property or physical assets, then money must be borrowed from abroad or else the domestic stock of foreign-exchange reserves or gold must be decreased. Traditionally these courses of action were considered a serious matter, because they meant that the country in question was consuming more than it was producing and the time had come to adjust local consumption in general, and foreign purchases in particular, in a downward direction.

Basically this is still the case, except governments no longer feel the same urgency to take corrective steps, since they reason that the necessary adjustment might take place automatically through the depreciation of the domestic currency. For example, if with an exchange rate of $1 for 2 German marks the United States were buying too much from Germany in relation to what the United States was selling, then theoretically under flexible exchange rates the dollar would *depreciate* (or the mark would *appreciate*), and eventually $1 would sell for less than 2 marks (for example, 1.75 marks). Under these circumstances, with everything else remaining the same, U.S. goods would be cheaper to Germans, and German goods would be more expensive for people in the United States. With people in the United

States buying fewer German products and Germans buying more U.S. products, the foreign account of the United States vis-à-vis Germany would be brought back into balance, with sales equal (or approximately equal) to purchases. This argument, which seems beautiful in its simplicity and which in a textbook world would be almost irresistible, eventually was accepted by otherwise sensible men as a result of the insistence of a few economists that a free-currency market was a logical extension of a free-market economy. Thus both the currency and bullion markets have at times, resembled bedlam.

At present many well-known private bankers, a majority of central bankers, and a number of so-called conservative economists such as Martin Feldstein and Arthur Laffer are calling for a retreat from the flexible-rate system to more stable rates. The reason is that in a world of speculators, where in the present context speculators specifically means gamblers, the world's currency markets have been turned into free-for-alls and casinos where the losers are not only some of the speculators, but also innocent bystanders watching the game. For instance, in the example just presented, the change in the value of the dollar relative to that of the mark came about *not* for the reasons given in the example, but because speculators, somewhere, were able to borrow large amounts of dollars (for example, on the Euromarket) and sell them for marks, thereby driving down the value of the dollar even if people in the United States actually selling more goods and services to the Germans than they were buying from them. To take an example cited by the former chancellor of the British Exchequer, Denis Healy, the Swiss franc has, on occasion, appreciated relative to the German mark (and other currencies) when there was no objective reason for it to do so, as a result of speculation. One of the apparent results of this situation is that small shopkeepers in the border areas and the employees of large hotels in the Swiss mountains have the satisfaction of knowing that many of their former customers are now shopping in Germany and skiing in France.

The place of gold is also important in this discussion because in the "old days" gold functioned at the ultimate unit of account. True, the dollar was also an official reserve asset in the sense that foreign debts always could be settled with dollars. But at the same time, dollars were *convertible* in that foreign central banks, if they felt that their holdings of dollars were too large in relation to their total portfolio of official reserves (gold plus dollars plus the special drawing rights of the International Monetary Fund), could send some of these dollars to the United States in return for gold. Eventually, of course, the United States was forced to suspend convertibility because the number of dollars in circulation expanded to several times the U.S. gold reserves. Still, in the twenty or so years before the suspension of convertibility and the gradual introduction of flexible exchange rates, the United States had managed its external economic affairs in such a way as to make the dollar one of the foundations of the present remarkable system of international trade and finance. For this reason, among others, the economists mentioned above as well as a number of prominent bankers want a return to gold-dollar convertibility; and people such as Eugene Birnbaum, chairman of the International Monetary Fund advisory board, want even more—fixed exchange rates based on an established price for gold. I believe that it is too soon to be thinking about a reintroduction of either convertibility or fixed rates; the best that can be done now is to take steps to have the dollar supplemented in its reserve role to an increasing extent by other hard currencies. But the present arrangement must definitely be modified.

A caveat is necessary at this point. The sudden demand shifts brought about by the change in oil prices did, in fact, require very rapid and frequent exchange-rate changes. However, I believe that in theory these could have been brought about by a "supreme court" of economists and upper-echelon civil servants meeting every monday morning (just as the U.S. Federal Reserve Board's open-market committee meets) and, if necessary, altering the exchange rate (just as the money supply or interest rate is altered). Moreover, as is well known by now, demand management in the wake of exchange-rate changes is as important in external adjustment as the exchange-rate changes themselves, since in order to bring about improvements in the balance of payments, resources usually must shift from the domestic sector to the export sector, particularly if imports are insensitive to changes in import prices in the short run. Shifts of this nature, as well as other policy measures, should be facilitated if exchange rates are being set systematically.

Two more topics should be mentioned in this introduction: inflation and the Euromarket. Inflation needs no explanation, although there does seem to be some question as to what to do about it. The simple (though perhaps incomplete) answer is to bring wages and salaries into line with productivity. One well-known recipe telling how this should be done focuses on controlling the supply of money. Roughly, if wages and salaries are increased without a corresponding increase in productivity, then producers will have to raise the price of their products in order to cover these increased costs. But in order for people to purchase these products at the new higher prices, the total amount of money in the economy will have to be increased. Thus, by increasing the amount of money, the authorities "ratify" unjustified increases in earnings. Conversely, by not increasing the quantity of money, the authorities force employers to refuse unjustified wage and salary demands, since consumers will not have enough money to purchase their output at the new higher prices. When output is not sold, both prices and production have to decrease, which generally causes unemployment of the people producing this output. Eventually, or so the story goes, employees get the message and stop asking for higher wages and salaries. As far as I can tell, this story is only partially true, since huge numbers of unproductive citizens have managed to insulate themselves quite well from the vagaries of the money supply.

One thing must be emphasized here: Productivity is *not* necessarily tied to willingness to work hard. Many people are working distinctly harder today than they were a few years ago; but their productivity is not increasing a great deal because often productivity is a function of the kind and quality of the machines at the employees' disposal or, equally important, the kind and quality of the machines being used by the people with whom employees cooperate or even society as a whole. As was emphasized in chapter 2, the rising price of energy is going to have a negative effect on the amount of equipment which will be available for the labor force; as a result, the employees' spending power is going to decline regardless of what they do—short of establishing another energy technology. Already this trend can be seen in the United States, where declining productivity, as well as the need to pay an increasing amount of real output for imported oil, should ensure the continuing descent of both the living standards and the economic prospects of that country, over the near future.

What is needed to turn this situation around is a radical attack on productivity and perhaps tastes. Welfare should, and could, be replaced by subsidized employment in both private and public sectors. Also, in the payment of unemployment compensation, a clear distinction must be made between individuals with work records and newcomers to the labor force, in the sense that the latter are *not* given the prerogative to refuse offered work because they consider it unsuitable. More emphasis must be placed on mechanical skills and physical training in elementary and secondary schools; and in universities, particularly in the social science faculties, there should be an across-the-board downgrading of research and an upgrading of teaching. Very large cash incentives (free of tax) should be available to individuals and corporations for significant innovations in crucial fields (for example, energy and medicine). Satisfactory performance on the part of university students and people in technical schools should be generously rewarded as a matter of course, with performance measured by centrally administered examinations. There should be a general emphasis on preventive medicine, to include voluntary participation in physical exercise during working hours at full pay; more important, governments should establish fitness centers at the neighborhood level; and, last but not least, priority must be given to integrating the socially and physically handicapped into productive employment.

In case some of these suggestions appear peculiar, the reader should remember that in the post-World War I period Europe's greatest economist, Joseph Schumpeter, was appointed finance minister of Austria for the explicit purpose of stopping the Austrian hyperinflation. Needless to say, he failed, although he had a complete arsenal of conventional wisdom at his disposal. One of his conclusions concerning this experience was that stopping inflation was a question of political will rather than economic theory and that eventually democracy itself would be destroyed by inflation because politicians would not dare to oppose it. Now history has provided us with a long list of smashed democracies, but as far as I can tell, the principal explanation in almost all cases is the presence of people like Schumpeter close to the levers of power—people with an elephantine capacity for ignoring real, but solvable, problems in favor of intellectual nuances. However, Shumpeter's successor, Monsignor Seipel, stopped the Austrian inflation cold; and although he was not in the hearts-and-minds business, he did not find it necessary to introduce any extremist political concepts into that country's everyday life.

One final point remains to be made on this subject—perhaps the most important of this chapter. The world economy may now be in a situation where a high level of inflation is indispensable. As more real resources are turned over to OPEC in return for oil, there is going to be less for the industrial world. However, the global redistribution of *real incomes* is concealed to a certain extent by maintaining or increasing *money incomes* (and thereby prices) in the industrial world, which perhaps obviates certain dramatic political reactions that such a redistribution would normally evoke. Similarly, perhaps high inflation rates have contributed to the smooth functioning of the international financial system.

If we consider the recycling of petrodollars, which is a transfer of a portion of the income of oil producers through various financial intermediaries to borrowers in the industrial world and the Third World, it seems clear that large amounts of money will not be supplied to the financial markets unless interest rates are fairly high. Instead, oil which is essential to the functioning of the industrial world would be left in the ground. As it happens, though, interest rates *are* high because the *money interest rate,* which is what lenders generally look at, contains an inflationary premium that boosts it over the *real interest rate.* For example, in a country in which the price level is unity, if a man has \$100, he can buy 100/1 = 100 units of real goods. And if he saves the \$100 at an interest rate of 10 percent and there is no inflation, then after a year he has \$110 and can buy 110 units of real goods. But if there is a 4 percent rate of inflation, then the price level is 1.04 after one year, and when this man get \$110 from the bank, he can only buy 110/1.4 = 106 units of real goods. Thus although the money rate of interest is 10 percent, the real rate of interest is only 6 percent. The difference between these, which is called an *inflation premium,* is 4 percent. In some countries, in fact, the real interest rate is negative, although this situation is disguised by high money rates. (Without going too deeply into this matter, I think it should be clear that *if* money interest rates are low and oil is left in the ground, then the oil price will rise. This price rise will increase world inflation because oil is an important input in consumer and industrial goods and when its price rises, the price of these goods is increased. In turn, this will drive up money interest rates.)

What about the other side of the financial market? It seems equally clear that large amounts of money will not be borrowed, at high interest rates, to finance projects having a low yield when measured in real terms. What happens, though, is that these high interest rates become, effectively, low rates because, as a result of inflation, the borrowed money often can be paid back in a *depreciated* currency, in the sense that the outputs from these projects sell at inflated prices.

We close this section by introducing the Euromarket. In the 1950s several Eastern European countries desired to keep money in the "capitalist" world in order to finance trade. Because of the cold war there was some apprehension about using U.S. banks, and so a large quantity of dollars were deposited with the Banque Commerciale pour l'Europe du Nord in France. The bank's cable and telex address was Eurobank, and brokers began to refer to the dollars deposited there as Eurodollars. Dollars deposited abroad have since been labeled Eurodollars, and the appellation *Eurocurrencies* has been attached to other currencies deposited outside their home country. An equally important phenomenon was the sterling crisis of the 1950s. During this episode, United Kingdom authorities put severe restrictions on the borrowing of sterling by nonresidents as well as on lending by British banks. This caused banks in the United Kingdom to turn to dollars as a substitute for sterling. Officially, these dollars were short-term assets that had been deposited in European banks without having been converted to local currencies. Moreover, the dollars were available because certain changes in U.S. bank regulations made it advantageous for individuals and firms to maintain large dollar deposits outside the United States. Initially this market was known as the Euromarket. Concomitantly, a Eurocurrency deposit is a deposit in a currency other than that of the country where the bank is located. At present the term *Eurocurrency* covers all expatriate funds involved in foreign dealings, including transactions arranged through financial centers in Southeast Asia and the Caribbean. The most important Eurocurrencies are the German mark and the Swiss franc.

London is the center of the Euromarket; in 1979, 256 banks in that city were authorized to deal in Eurocurrencies, with about 40 percent of this business being handled by U.S. institutions. One reason is the capacity of the London money market as compared to those of the Continent. It is still possible to purchase bonds worth 10 million British pounds in London in a few minutes without anyone raising an eyebrow, while in Frankfurt or Zurich this single transaction would probably move the price. Because most national governments are very strict about regulations concerning the operation of national banks in the home country, while being somewhat more liberal toward foreign banks operating in the domestic market and, even more so, toward local banks operating abroad, the Euromarket has proved extremely popular in the banking community. Some reservations are now being expressed about this market, however, because of the growth in its size. In 1967 the Euromarket held some $18 billion; today its gross size is approaching $1 trillion, as a result of the enormous U.S. oil deficits that are, to a considerable extent, being paid in newly manufactured dollars.

Eurodollars also accounted for 65 percent of the total U.S. money supply in 1978, as compared to 20 percent in 1970, which means, among other things, that this may be the first time in history when a country has allowed so large a share of its legal tender to come under the control of foreigners. Still, it could be argued that the Euromarket has proved invaluable for recycling oil revenues, since the international financial system might have collapsed if a large fraction of the billions of dollars in oil revenues generated since 1973 had gone directly into the currency markets. Equally important, the increase in oil prices had, in itself, a deflationary effect on the world economy; in the absence of effective demand management by governments in the industrial countries, it was essential to return to the normal channels of trade and investment that part of oil revenues which could not be spent by the oil-producing countries.

The Recycling of Petrodollars

Since the recycling of petrodollars is discussed in the beginning of this chapter, the next step is to inform the reader that there is nothing more complicated about this process than any individual putting a portion of her or his monthly paycheck into a bank and the bank lending this money to whomever they consider to be worthy borrowers. The problems that arise in the case of petrodollars involve the size of potential deposits and their ability to influence the international economy.

In Banks (1979a) a certain amount of attention was directed to some comments about the problems of recycling made by Bell (1976). As things have turned out, one of my points of disagreement with Bell must be rescinded. Bell questioned the ability of financial institutions to continually absorb large additional inflows of deposits. In one of my earlier books, I denied the validity of his observation, given the possibility of such things as negative interest rates and the potential existing at that time for directing a larger portion of oil revenues into highly liquid, short-term bills and blue-chip securities and property. But now it appears that, given the new huge boost in oil revenues caused by the recent oil price increases, the international banking fraternity is rapidly approaching a point where it cannot possibly handle these revenues at normal rates of interest because there are not enough high-yield projects in the world that can absorb these funds.

It could be suggested, though, that this is a short-term problem, since if the international financial system cannot accommodate these revenues with the existing structure of prices, interest rates, yields, and so on, then either these parameters will change or, over time, oil revenues will change—or both. For example, as was pointed out in the last section, oil production might decrease, which would cause inflation rates to rise, which in turn would increase the inflation premium on interest rates (and thus cause money interest rates to climb), which might make oil producers willing to maintain high outputs since they can lend their revenues at what they think is a reasonable rate of return. By the same token, sponsors of low-yield projects, faced with high interest rates, might still be interested in borrowing money since the inflation means that they can count on higher prices for the output of their projects, and in some cases they can gain by making their repayments in a depreciated currency.

In the United States at the time this book is being written the high rate of interest and high inflation rate have clearly not shaken business confidence although this situation could change overnight. The reason is that, both private consumers and firms have developed an inflationary mentality and are borrowing as much as they can. Even so, with the average inflation rate for the industrial world now at 15 percent, financial circles are more apprehensive than ever, reasoning that regardless of short-run gain, no good has ever come out of a state of affairs where money fell in value at its present rate. Also it should not be forgotten that as time goes by, oil producers might become more disposed than ever to leave a larger part of their oil in the ground regardless of the temptation to extract it and invest the revenues in financial assets offering record yields, because in some of these countries there are people who have started to ask themselves, and also their friends, just what economic prospects they would face if this commodity started to run out. In that case, the industrial world would have to learn to live with both high inflation rates and a depressed level of industrial activity, with the latter probably involving even higher levels of unemployment than those being experienced at present.

Another dilemma in the recycling of petrodollars involves loans to the Third World. These countries apparently tapped the international capital markets for about $230 billion between 1974 and 1979, bringing their total debt to over $350 billion. Another $80 billion or so may be added in 1980, and as is becoming increasingly well known, particularly in the nonbanking community, much of this money has been borrowed for projects that should never have been considered, much less started. Some figures on LDC debt are given in table 6-1. In connection with this table, it should be mentioned that the Federal Reserve Board has begun to remind U.S. banks,

Table 6-1
Estimated Medium- and Long-Term Debt of LDCs
(billions of current U.S. dollars)

Debt	*1970*	*1977*	*1985*	*1990*
To Private Creditors				
Low-income countries	2	10	16	19
Middle-income countries	30	145	422	752
	32	155	438	771
To Official Creditors[a]				
Low-income countries	15	30	108	183
Middle-income countries	22	66	194	324
	37	96	302	507
Total	69	251	740	1,278
Total at 1975 Prices	113	231	348	449

Source: World Bank documents.
[a]Includes multilateral.

publicly, about the dangers of excessive lending to the Third World. The basic issue here is that many Third World countries are caught in what has been termed a *debt trap*: They must borrow an increasing amount in order to service past debts, although these debts have little relevance to the basic economic needs of the countries doing this borrowing. As far as I can tell, this cycle can come to an end only with a major war and/or the crash of the interational financial system.

A number of important personalities and insiders have begun to insist that the International Monetary Fund (IMF) and World Bank should play a greater role in the petrodollar drama. Just what role these overrated institutions are capable of handling is an interesting question. Perhaps if the internal economies of some of the Third World countries could be improved, then OPEC surpluses would have a natural home; but the improvement of these economies is strictly an internal political and sociological matter. It is highly unlikely that any serious governments in these countries would put their economic fate in the hands of the IMF or World Bank.

Some prominent bankers feel that the solution to the recycling problem lies in new financial instruments. But no financial remedies can compensate, in the long run, for the effect on the world economy of actual or potential energy shortages which keep highly efficient societies such as those of Japan, Germany, and Sweden from fully utilizing their advanced technical proficiency. By way of clarification let me explain that what these bankers want is a new bond, certificate, or something similar whose return is guaranteed in real terms, and thus is attractive to OPEC lenders.

One of the emerging superstars on the international monetary scene, Rimmer de Vries of Morgan Guaranty Trust, has come out in favor of minor "off-market" measures in order to avoid the danger of import and capital controls, which he sees, and correctly, as being a possibility once the new wave of OPEC surpluses hits the exchange markets. One of his suggestions involves the direct channeling of OPEC funds to central banks and to the IMF through these banks. This comes down to a program for getting these funds out of circulation for as long a period as possible. The mechanics of this arrangement are as follows.

Oil producers could sell dollars to, for example, the West German Bundesbank in return for securities denominated in marks. The rate of interest on these securities would be somewhat lower than the market rate, which would be the Bundesbank's compensation for the exchange risks involved in accepting dollars. Then the Bundesbank would invest these dollars in U.S. government securities at terms above the market rate, thereby passing along some of the remaining exchange risks. All this means is that, given the number of dollars which will accrue to the sellers of oil in the next few years, this measure has no future. Eventually (and perhaps to begin with), Germany would be unwilling to buy these dollars unless it were in return for bonds with a very low rate of interest, and at the same time the interest rate on U.S. securities were very high—so low and so high, in fact, that neither the United States nor the OPEC countries would be interested in the first place. Therefore the necessary adjustment to the developing economic situation will have to be made *in* the market, and not *off* the market; and as de Vries has correctly indicated, it will not be an adjustment to our liking. For what it is worth, however, there is one consolation. In the short run, this adjustment and ensuing ones are unavoidable.

Since controls were mentioned above it should be noted that if the Euromarket continues to expand at its present rate, some attempt will undoubtedly be made to limit its prerogatives. Up to now the most significant attempt to interfere with the Euromarket was the agreement by the Group of Ten (industrial countries) and Switzerland not to deposit their foreign exchange reserves in the market. This measure was abrogated during the liquidity crisis that followed the collapse of the Herstatt bank in Germany, but subequently reimposed with the liquidity of the Euromarket attained previous magnitudes. The present talk though is of direct restraints on the expansion of credit, and apparently a number of governments, to include those of the United States and Britain, are sympathetic to this point of view.

There is also some talk of creating a rival banking center in the United States which would be easier to monitor. Ostensibly this would come about if the Euromarket did not begin to behave itself. The question that comes into my mind upon contemplating all this concerns the motives of the individuals who are encouraged, or encourage themselves, to hatch and elaborate these schemes. It is probably true, of course, that some of them believe themselves to be rendering a service to humanity; but it so happens that no game playing with bank locations can alleviate the real difficulties that the industrial world is going to face because of energy shortages and insufficient capital formation.

Apparently some of the socialist countries are also in the market for capitalist lucre. Poland and Hungary are probably the most active of these countries. Total East European debt to eurobanks is about 50 billion dollars, and much of this has been used to finance the import of western technology. It also appears that Russia has been scaling down its borrowing because the appreciation in the price of gold (of which it is a major producer) has provided that country with a sufficient supply of hard currencies. In fact the Russians have been repaying their debts ahead of time, and it has been suggested that if Russia runs into any technological knots in connection with the augmentation of its energy supply, it now has enough money to hire the best expertise in the capitalist world to help untie them—assuming that outside help is needed. The truth of this contention can be glimpsed in the light of the present U.S. wheat boycott of Russia, which the Soviet Union is in the process of surmounting by simply increasing their purchases from other sources, employing their ample reserves of gold and foreign exchange.

Oil and Gold

There is a tendency for a great deal of oil money, or money influenced by oil, to be directed into the commodities markets, in particular into the purchasing and/or holding of minerals and metals. The most important commodity in this respect right now is gold. Although it is impossible to know just how much gold is held by private individuals, it is thought to amount to at least as much as is being held by the world's central banks and monetary authorities. This latter figure comes to approximately 32,000 metric tons (t) or 1.2 billion ounces (Goz). [If we look at flows, the total supply to the private market during 1979 was about 2,000 t, or 70 Moz. Some 720 originated in South Africa; 400 t, in the centrally planned countries, principally the Soviet Union; and 550 is the result of the sales of monetary gold by the United States and the IMF. These sales, whose ostensible purpose was to take the pressure off the gold market, should never have taken place, as was pointed out in Banks (1979a).] In the case of the United States these sales amounted to the government giving away billions of dollars.

The value of the above monetary gold at present is approximately $700 billion. There is some controversy about the rate at which various monetary authorities have been increasing their holdings of gold. There have been claims by insiders such as the well-known international economist Edward M. Bernstein, Robert Guy of Rothschild & Sons, and Michele Sindona that certain central banks have been ditching their dollars in favor of gold. However, the central bank of Singapore is openly in the market, and thirteen governments of LDCs have placed reserve bids with the IMF which guarantee them the right to acquire gold at the average bid price prevailing at one of the IMF's monthly gold auctions.

A considerable amount of gold being bought by private persons is financed by money borrowed in the Euromarket; needless to say, more than one fortune has been made by resorting to this procedure. Just who are the large private buyers of gold is difficult to say (although Merril Lynch International estimates that U.S. citizens own about 3,000 tons.) Certainly a large part of the $10 billion or so that left the Gulf region during the Iranian revolution moved into gold, either directly or indirectly; and apparently some "metals merchants" are making daily purchases for Middle Eastern buyers, which puts a relentless upward pressure on the price of gold.

During 1979 the *average* price for gold was about $300 per ounce, while at present the market price is close to $550 per ounce. (In January 1980, however, the price reached $875 an ounce.) The only thing that keeps this price from going higher (which it will probably do, despite occasional falls) is the knowledge by many present and potential buyers that gold's official price was maintained at $35 per ounce from 1934 to 1971 while the average price of everything else tripled in tune with a growing political instability, and that there is a basic irrationality in this procedure of digging gold out of the ground at great expense in order to transfer it to the interiors of bank vaults and mattresses. The value of gold obviously rests on a mythology, but in a certain sense the same is true of the value of paper money: the myth in the latter case is that governments, and the economists who advise them or manage their monetary affairs, are always capable of, or even interested in, comprehending the myriad social and economic factors that lie behind maintaining a stable currency. The United States may also have entered into some political agreements with various other governments regarding the disposition of the dollars in the possession of these governments: The Western European allies of the United States (with the possible exception of France) have been "warned" on more than one occasion not to decrease their stocks of dollars too rapidly; and it has been suggested that some OPEC countries have reached an understanding with the United States on this matter. Last, but not least, the prime rate of interest in the United States is close to 15 percent. Many people who, a few months ago, regarded the dollar as a pariah, now want dollars in order to purchase U.S. securities and bank accounts; and as the value of the dollar rises relative to that of the foreign currencies that are offered for them, there is also a tendency for it to rise relative to gold. But the reader should understand that the political and sociological realities of U.S. life, as well as the increasing cost of oil imports, *and* the huge potential supply of dollars in private and official inventories around the world will ensure that the greenback's present respite is brief. (At one point, in 1979, the prime rate touched 20 percent.)

A few words can now be said about the recent history of gold, beginning with the conference at Bretton Woods (New Hampshire) in 1944. Under the so-called Bretton Woods system, which was established at that conference, the dollar and pound sterling, and gold were established as legal reserves for the international monetary system proposed for the postwar period. (The pound, which was initially a more important reserve currency than the dollar, was later dropped.) Lord Keynes, who was the star of the conference, wanted gold abolished from international monetary affairs and a kind of international central bank set up which would supply, in the form of something to be called "bancor," the reserves needed to promote international trade. Fortunately this advice was ignored. In the gold-exchange regime that was introduced, gold and the dollar (and initially the pound) were placed on the same level, with the dollar fully convertible into gold. In fact, the dollar was even more valuable than gold because it could be invested in interest-bearing assets whereas the price of gold was fixed (at $35/oz) and consequently its purchasing power depreciated relative to that of paper currencies.

Until the Vietnam war, the Bretton Woods system functioned to the satisfaction of almost everyone. World trade and the currency markets were gradually liberalized, and full convertibility among the currencies of most of the industrial countries of Western Europe and North America prevailed by 1958. Although a number of currencies were devaluated against the dollar in 1949, exchange rates generally were stable during the 1950s, despite speculation in the revaluation of the German mark and devaluation of the pound. Following the revaluation of the mark and of the Dutch florin in 1961, the exchange rates of the major currencies remained stable until 1967. But the devaluation of the pound in 1967 and the financing of a large part of the Vietnam war with newly printed money instead of taxes introduced an unsettling note into the system. Eventually the sheer quantity of dollars pumped into the world economy raised the rather embarrassing question of just what degree of convertibility the United States would or could support, given its limited stock of gold. President Nixon finally clarified this issue by abolishing gold-dollar convertibility, and the world went on a dollar standard.

Now, under a dollar standard, accelerated monetary expansion in the United States has a tendency to lead to monetary expansion everywhere. To see how this works, let us assume that dollars are "created" in the United States in order to permit people in that country to purchase foreign oil. Many of these dollars are then spent in some Western European country by the sellers of oil. This creates an excess supply of dollars in that country (which probably includes a downward pressure on the price of the dollar in the foreign-exchange market); and as these dollars find their way into the central bank of the country in question in return for local currency, the money supply of that country is increased. (Remember that the most common definition of the money supply of a country is currency in the hands of the public plus demand deposits. Money thus defined is usually called M_1.)

Since governments are now free to allow their exchange rates to move up and down in response to a change in the supply and demand of currencies, theoretically they are insulated from the above phenomenon to some extent, since the excess supply of dollars would either reduce the value of the dollar or increase the value of local currency so as to greatly increase the demand for dollars to be used for making purchases in the United States. But, as I explained earlier, a sane government does not make a point of allowing the currency markets to determine the value of its money if hardships will be imposed on important sectors of its economy. Both the Japanese and the Swiss have reacted strongly in the past year to prevent the value of their currencies from appreciating to a point where many of the products and services that they sell to foreigners would be priced out of the market. This has helped convince many dollar-holders desiring other assets that they would be better off moving into gold than into yen, swiss francs, or other currencies. In turn, this has helped depress the value of the dollar relative to that of gold and has led to charges by some OPEC money managers that the market gold is being rigged to reduce the purchasing power of oil, even though much of this gold is apparently being purchased by individuals, or perhaps even governments, in the OPEC community. By using the analysis just developed, however, one could argue that if the purchasing power directed toward gold had been allowed to "bid up" the value of other currencies vis-à-vis the dollar, then perhaps the price of gold would not have escalated, but the purchasing power of oil still would have deteriorated. In either case there would have been an increasing pressure from OPEC's side to raise the price of oil.

The increase in the price of gold may also have an inflationary bias. The money value of consumer spending could be boosted by a so-called wealth effect that is due to the upvaluing of that part of consumer assets represented by holdings of gold. Moreover, the gold reserve of many governments has increased drastically, and now the value of gold held in official reserves exceeds the amount of paper money used for this purpose. This makes governments feel wealthier and enables them to believe that they possess the wealth to support a higher level of official spending, borrowing, or money creation. It has also been claimed that the gold owned by countries will determine the terms on which they can borrow money. On the plane of the individual or the firm, when the price of gold rises rapidly and gives some indication that it will rise again at an equally spectacular rate, money that would normally flow into productive channels is used to finance gold hoards instead. Hence this causes more inflation because the goods that would normally flow from these productive investments do not come into being. Many individuals who are not lucky enough to own gold in one form or another become demoralized when they realize that those who were lucky are in line now to realize windfall gains of the type that were once experienced only in television serials and by no means could be obtained as part of the remuneration for conventional employment. These people respond by insisting on higher salaries or wages.

Next we look at the pattern of gold supply and demand for 1975-1978, remembering that earlier gold was being absorbed by official sources instead of being sold. (See table 6-2).

One of the peculiarities of gold is the lack of a common standard for valuing existing official reserves. On one hand, Britain and the United States value their gold stocks at $42.32/oz whereas Canada and Portugal

Table 6-2
The World Supply of and Demand for Gold, 1975 to 1978
(*t*)

	1975	*1976*	*1977*	*1978*
Supply				
South Africa	713	713	700	708
Canada	51	52	53	52
Latin America	39	45	47	47
United States	33	33	34	30
Papua, New Guinea	19	20	23	24
Other	108	108	110	110
Total	963	971	967	971
Communist (Mostly Russia)	149	350	450	450
IMF Sales			184	184
U.S. Treasury			—	126
Total Western Official	15	155	195	325
Total	1,127	1,476	1,796	2,056
Use				
Jewelry	516	932	979	1,001
Total Industrial (Jewelry plus Other)	697	1,142	1,201	1,248
Investment (Hoarding)	424	311	406	493
Total[a]	1,121	1,453	1,607	1,741

Source: South African and Australian official documents; the Bank for International Settlements; IMF; and Credit Suisse.

[a] *Total*, in this case, means total industrial plus investment. Note also that there is a slight discrepancy between total supply and total use.

value theirs at the former official price of the IMF, or 35 special drawing rights per ounce, which at present translates to a price of $45/oz. The French, on the other hand, value their official reserves at the market price, and the Japanese, who do not reveal how much gold they own, value gold at either the price at which it was bought or the free-market price, whichever is lower. The advantage of one arrangement in comparison to another is unclear to me unless some of these authorities believe there is a direct link between the value of a country's official gold holdings and its money supply, which is theoretically the case in the event of a perfectly neutral attitude by a country's monetary authorities. This link is weakened by keeping the official price of gold independent of the market price.

Futures markets for gold are now available in some countries. By late 1979 the volume of gold being traded on U.S. futures exchanges came to 4 Moz/d. Some U.S. banks are also issuing gold certificates which permit the purchase of gold in small amounts and various gold bullion coins are also available. Arrangements of this type tend to increase demand, but there is some question as to whether supply can keep pace. Production in South Africa has been declining since the early 1960s, and Caldwell (1980) has suggested that by 1995 the output of that country may be down to one-half its present level. It also appears that the intensified level of exploration stimulated by the rapid rise in the price of gold has not resulted in any important new finds. These factors help to explain the attractiveness of shares in gold-mining companies. Many South African gold-mining shares yield over 20 percent a year, and their yield has not been materially influenced by ups and downs in the gold price.

It is now being suggested in influential circles that the renewed strength of gold makes it a candidate to back a new IMF *substitution account* capable of absorbing the enormous number of dollars in the world's currency markets. This would involve a new issue of special drawing rights (SDRs), those reserve assets for governments and central banks that are issued by the IMF. (To a certain extent, they were what Keynes had in mind when he suggested a paper reserve asset which would be issued by an international central bank. He did not, however, see this bank as being a highly politicized institution directed by international bureaucrats of the type associated with many agencies of the United Nations.) In 1978 these SDRs amounted to almost 3 percent of the world's reserve assets, and in theory they are on the same level as gold, dollars, or any other currencies that might come to be held in reserve portfolios. This scheme thus calls for the substitution of an indexed value—the SDR—for the depreciating value of dollar balances held by oil producers. As it happens, though, the dollar carries a large weight in determining the value of an SDR; and should the dollar crash, the SDR would become an extremely unattractive store of value. (Just now there are about 13 billion SDRs outstanding, and these are worth about $17 billion. Another 8 billion are authorized for future issue.)

Another artificial currency is the European Unit of Account (EUA), which was originally adopted by members of the European Economic Community (EEC) as the common accounting measure. At one of the European summit meetings, it was decided to begin construction of a new European monetary system, with the EUA graduating to the role of an embryonic European reserve currency. It would, however, be renamed the ECU (which is a name related to the ancient French gold coin, the *écu*). Then national currencies of EEC members would be required to maintain values in relation to the ECU which would, if all went according to plan, keep them from straying too far from one another. Members of this organization would be expected to contribute to a pool of about $25 billion in reserves and to match this contribution with a similar donation of national currencies. Thus the total pool of reserves of U.S. dollars, gold, marks, French and Belgium francs, sterling, lire, guilders, and Danish crowns would amount to approximately $50 billion, which exceeds the total reserves held by the IMF. These resources would be administered by representatives of the central banks of the EEC countries to begin with. Intervention in the currency markets would involve both these pooled reserves and the resources at the disposal of national central banks, in order to maintain the system's currencies within the agreed limits.

Member countries would have access to credits in ECUs if they experienced balance-of-payments disequilibria; however, these credits would be subject to rigorous control. The European monetary system, although it has larger resources and is concerned with fewer currencies, is very similar to the Bretton Woods system. But one wonders whether it could actually manage exchange rates given the tremendous speculative forces that are occasionally brought to bear against the currency markets. Even so, some obvious economic distortions caused by speculation could be ironed out, since in time some provision is to be made for at least some harmonizing of national economic policies; and the most powerful unit in this arrangement, the Germans, are evidently prepared to do everything possible to make it succeed.

Oil and the U.S. Dollar

We are now ready to examine a topic of major importance—the relationship between the price of oil and the value of the U.S. dollar. The hypothesis advanced is that the change in the value of U.S. currency prior to late 1979 was a direct result of the drastic increase in the dollar value of U.S. oil imports. This may seem obvious to many readers, but David Rockefeller of the Chase Manhattan Bank, Robert Solomon of the Brookings Institution, and Irving Kristol of New York University and the *Wall Street Journal* take the position that the dollar has become a problem because the high rate of inflation of the United States relative to those of its trading partners has caused both foreigners and people from the United States to shy away from dollars and dollar denominated assets. In addition, U.S. exports have become less competitive. All this is true, of course, but secondary.

We can begin the analysis by recalling that oil is used as an input for almost all the goods and services produced in a typical industrial country. Thus the oil price rises of 1973-1974 put an upward pressure on the general price level of the oil-using countries. If we take the case of the United States, we have the situation shown in figure 6-1. The oil-price rise is a cost increase which translates into a shift of the aggregate supply curve from S' to S'': At every value of the market price, less is offered since it costs more to produce each unit of output; with an unchanged demand curve, the aggregate price level increases from price p' to p''. (Furthermore, if we make the assumption made in 1974-1975 that a large part of the investible surplus of OPEC would be spent or invested in the United States, then, as shown by the shift in the demand curve to D^*, the U.S. price level is increased further to p^*.) Note also the decline in production to q'' or to q.*

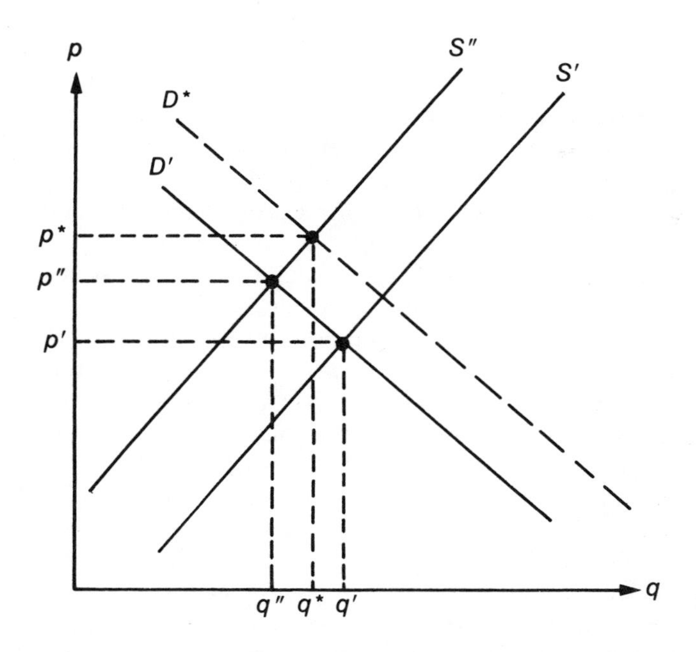

Figure 6-1. Effect of the 1973-1974 Oil-Price Rises on the U.S. Economy

Next we examine the effect of just these events on the value of the U.S. dollar. The increased demand for U.S. goods by OPEC, in causing the U.S. price level to increase relative to that of other industrial countries, would cause a depreciation in the value of the dollar, *ceteris paribus. But* since the other oil-importing countries require dollars in order to pay for their oil, and since these dollars must be obtained from the United States, the value of the currency of the other oil-importing countries will depreciate relative to the dollar as they attempt to raise their exports to the United States. Also many of the dollars earned by OPEC would be coming back to the United States rather than being offered for foreign currencies or gold. This gives us two opposing tendencies, and as we shall see, the world financial community initially expected them to be resolved in favor of the United States. Later there was a change in opinion.

In the period immediately after the first oil-price rises, until February 1974, the dollar appreciated relative to most currencies. For example, it rose by 11 percent versus the West German mark. Also in this period world inflation rates increased relative to price rises in the United States, and short-term dollar interest rates on the Euromarket declined, reflecting the market's expectation of lower interest rates in the United States vis-à-vis those of Japan and Western Europe. This was interpreted to mean that there would be a boom in the fixed-interest bond market in the United States which could carry over to all U.S. financial assets. In general it was expected that the health of the U.S. economy would improve radically by comparison to that of its energy poor trading partners, and the U.S. balance of payments would prosper accordingly.

Later, however, when it was perceived that the United States, with its already high rate of unemployment, would have to hold the line against a politically and socially unacceptable increase in unemployment (represented in figure 6-1 by the fall in production from q' to q'' or q^*) by monetary expansion, Eurodollar interest rates gradually climbed to record levels, reaching 14 percent in some cases. This increase was an inflation premium defined earlier in this chapter. In addition, by the autumn of 1974, the oil-producing countries were rapidly increasing their imports from Japan and Western Europe and exchanging their dollars for marks, yen, and so on, which tended to decrease the value of the dollar with respect to these currencies. The major surprise of this period was the ability of OPEC countries to successively increase their ability to absorb their oil revenues: By 1978 they were spending or lending (in both the Euromarket and the United States) all but a few billion of their oil earnings of $125 billion. But there was also a growing tendency for OPEC and other governments, as well as a number of financial institutions, to "diversify" out of dollars and into other currencies. This was caused by the suspicion, eventually growing into a conviction, that the United States could not or would not decrease its imports of oil.

We have now reached the crux of our discussion. The United States was, and is, financing its increasing oil imports by printing greater amounts of money, which then finds its way into a world that already has more dollars than are required. The reason why the United States has been printing this money is to allow U.S. firms to buy the energy they need to maintain production and employment, which also permits U.S. residents to maintain the energy-intensive lifestyles that are, to a certain extent, associated with an unchanging or advancing level of real income. At the same time, the governments and financial institutions in other countries are diversifying out of dollars, when they can, in order to drive down the price of the dollar and reduce the size of the dollar-denominated oil-import bills they will have to pay. This behavior contributes to maintaining their real incomes. This blend of efforts and accomplishments explains a large part of the high rate of inflation in the United States as well as the decline in the dollar. The initiating factor for both these torments is the rise in the oil price and the attempt to mitigate its effect on the U.S. economy.

North Sea Oil (and Gas)

The oil of the North Sea apparently saved Britain from an accelerated trip to pauperdom. I use the word apparently because, as a spokesperson for the British Trade Unions Council recently emphasized, if the present government of Britain insists on being guided by the economic philosophy of Milton Friedman, then that country will soon gravitate to the status of a banana republic. The philosophy in question is a kind of crude monetarism which relys on unemployment to cure inflation; but as things are working out, both unemployment *and* inflation are increasing. In addition, four research organizations have predicted that Britain is on a collision course with economic disaster. (Among these are the highly competent National Institute and the considerably less competent London Business School.) The reason is simple: In the past year industrial production in the United Kingdom has fallen by 4 percent. Even worse, investment may be collapsing.

The situation before the British government is clear. Unless care is taken, oil revenues from the North Sea will go into private consumption instead of capital investment and the improvement of primary and secondary education; or they will go into inefficient or unnecessary defense spending, charities associated with the Commonwealth and the United Nations, and so on. The fact is that these revenues must go into some form of investment because only the distortion in international resource allocation caused by the energy crisis, which works against the energy-poor countries of Western Europe and Japan and in favor of energy-independent Britain, permits a large portion of British industry to exist in the unrelenting rivalry which now characterizes much of the industrial world and which will become much more acute once Britain's competitors have reduced their dependence on foreign energy.

Still, changes for the better are possible, even at this late date. The best in British industrial practices and education is very good by international standards, and the importance of Britain's ample supply of domestic energy is liable to increase rather than decrease as the OPEC-controlled oil price is adjusted upward. Proved and probable reserves in the U.K. sector of the North Sea now comprise between 1,500 and 2,600 Mt, and estimates are that between 900 and 1,800 Mt will still be found. The money value of these proved and probable reserves is said to be approximately $805 billion, whereas the value of natural-gas reserves under British control is about one-half as much. Annual revenues from this oil should rise from 3 billion British pounds during 1979 to at least 15 billion British pounds in five years or so. The all-inclusive cost of this bonanza cannot be reckoned just now, but as was pointed out earlier, offshore oil tends to be high-priced oil. For example, $9 billion has been invested in their concession by the Shell-Exxon partnership. Something that cannot be measured in terms of money, however, is the self-sufficiency in energy that Britain will enjoy during most of the 1980s. It is an advantage not available to any other large industrial country except Russia.

One of the major "problems" associated with British oil is taxes—what kind and how much? [The same is true for Norwegian oil. For a discussion of both these countries and their methods of taxing oil revenues, see the excellent surveys of Kemp and Crichton (1979).] The original idea of the British government was to ensure that almost all the British sector of the North Sea would be exploited, and taxes were designed which made it highly profitable to tap all but the most marginal fields. However, because of the unforeseen rises in the price of oil, virtually no areas of the North Sea are economically unattractive. The minimum size for an economic operation has been halved from 150 to 200 Mbbl of reserves to 50 to 80 Mbbl, and seismic and drilling activity continues unabated. The British government anticipates a veritable tax bonanza, and it is being said in authoritative circles that the taxing of oil revenues will permit either a doubling of the defense or education budget or a spectacular reduction of the income tax. The Norwegian government, apparently, has already reaped such a bonanza.

The British will also permit large-scale investment in other energy technologies, and this is becoming more important than ever since soon there might be the danger that the generally poor performance of the British economy is disguished by the magnitude of oil revenues. The oil industry itself employs relatively few people; its contribution to GNP is modest; and the problems of world recession, inflation, and the comparatively low productivity of many sectors of British industry will exist long after the North Sea oil starts to run out. Accordingly, this oil must play its part in making Britain a viable economic unit in the next century. One way for this to fail is through encouraging the government to relax its overall supervision of North Sea activity. Briefly put, the duty of the government is to provide the oil companies with the economic incentives that will cause them to lift every barrel of recoverable oil in those fields. Undoubtedly this can be done through fiscal incentives rather than the wider private participation in the ownership of the oil fields espoused by some of the more avid partisans of the laissez-faire ethic.

Up to now only about 180 Mt of oil has been brought ashore from the British sector of the North Sea, and this is equal to approximately two years' consumption by that country. The oil firms operating in the North Sea apparently plan to export a certain amount of oil; but even within the conservative government there are influential voices which insist that the life of the oil fields must be maximized. It has been suggested that this could mean exports of no more than 100,000 bbl/d by 1985, although at that time these fields may have the capacity to produce 3 Mbbl/d, and domestic consumption will just exceed 2 Mbbl/d. The oil company approach to this matter is to argue that intervention to reduce production will reduce long-run income (and tax revenues). Just now, no one can provide any reliable information on the size of these exports.

Norway and Holland

The situation in Norway and Holland is also illuminating. Despite the advantage of possessing an impressive supply of domestic energy, these countries are facing certain economic difficulties, although Norway's are more imagined than real, since that country is obviously on its way to becoming the most prosperous in the industrial world, something even its unimaginative politicians will have a hard time preventing. Domestic demand in Norway is rising at a rate that is far in excess of the OECD average, which merely reflects rising per-capita consumption; and during the worst phase of the recession that hit the industrial world in the wake of the 1973-1974 oil price rises, unemployment in Norway was only about 1 percent.

Just how long the Norwegian oil will last, at present exploitation rates, is uncertain, since a large part of the Norwegian North Sea is unprospected. But intentions are to make it last at least one hundred years. In addition, a major gas find is in the process of being confirmed. Under the circumstance, Norway has a great deal of time to plan and begin constructing its future industrial structure. Some of this may be done with Swedish or German help in return for energy.

The Netherlands is not so lavishly blessed, but the Groningen gas field is perhaps the most important in the world in terms of the service it has performed for a single country: 85 percent of Dutch homes use gas for heating, and about 80 percent of the very efficient Dutch industry uses gas to one extent or another. Altogether, gas supplies more than one-half of the energy used in the Netherlands, and the fact is that with careful planning that country possesses a domestic energy base that is the envy of many of its neighbors. But at the same time, Holland is exporting large amounts of gas which it may need at home for prices that, to a considerable extent, are under world energy prices. In fact, the Dutch must now look forward to paying more for the gas they have begun to import (in order to preserve their own low-cost reserves from the Groningen field) than what they charge for exports.

This strange situation came about because of the huge sales made via long-term contracts many years ago when it was thought that when their gas ran out, inexpensive oil could always be imported in unlimited amounts. Between 1974 and 1978 import contracts for about 100,000 Mm^3 were concluded, and some of the long-term contracts referred to above were originally scheduled to run for twenty-five years. As things now stand, the Dutch government has threatened to cut off gas supplies to West Germany, France, Belgium, and Italy if the prices on these contracts are not renegotiated. There has also been an attempt to index gas prices to the price of oil, but apparently this arrangement has not been particularly successful.

Emphasis now is on conservation policies, which includes routing domestic gas to the most efficient users, more imports of natural gas (especially from Algeria and Norway), and a program to reduce energy inputs per unit of output by 10 percent by the year 1985 and a further 20 percent by the year 2000. A countrywide insulation program is also to be tried for the purpose of reducing household heating requirements. Other energy materials are also to be exploited, if possible, with emphasis placed on the very deep deposits of coal located in the Netherlands. These deposits, which amount to perhaps 100 Gt, are at depths of 1,000 to 6,000 m, and thus new technologies will be necessary to exploit them (such as *in situ* gasification and hydraulic mining). Interestingly, all this may represent a reversion to Holland's earlier energy policy. In 1955 coal contributed 96 percent of power-plant fuel in Holland, while in 1975 it contributed only 0.7 percent. The present intention is to gradually increase coal production and imports until they supply 40 percent of power-plant fuel by the turn of the century. Gas and oil, which presently supply 90 percent, will be restricted to 20 percent, while a decision on the composition of the other 40 percent will be made sometime in the early 1980s.

Final Observations

In the discussion of inflation presented earlier, nothing was said about the traditional method of slowing down accelerated price rises: —recession. The reason is that recession means unemployment, and unemployment levels in the industrial world have already reached calamitous proportions. The number of people in the industrial world who desire, but cannot obtain, work is approaching the 20 million mark. It will soon reach that milestone and continue to climb.

Obviously, when faced with this kind of situation, governments are unlikely to go around starting recessions, at least consciously. Yet I personally have encountered a number of well-tenured academic economists who argue that this is precisely what is required. Their reasoning originates from the following background. In the United States, Britain, West Germany, and Japan there have been twenty-seven slowdowns in economic growth since 1950, and in twenty-six of these the rates of inflation have significantly slowed. Of course, prior to 1973-1974, inflation in the United States usually meant an annual rate of aggreagate price increase greater than 2 percent. If the United States is lucky, inflation in that country during the 1980s might get down to 8 percent, depending on the energy situation; but for the rate to reattain 2 to 3 percent probably would require at least 12 percent of the labor force to be idle, as well as a depth and quality of social upheaval which would make the 1960s look like a strawberry festival. It is also true that more than a few economists also see recession as a cure for some of the ailments of the international financial system, since if recession does manage to reduce the U.S. price level, the value of U.S. currency might appreciate, and quite a few billion of the expatriate U.S. dollars which have been causing international bankers such anxiety would return to the homeland. In a notable seminar at Uppsala University in Sweden several years ago, a leading U.S. international economist estimated that 10 percent unemployment in the United States would go a long way toward restoring the credibility of the dollar. Needless to say, he did not anticipate having to make a direct contribution to the rehabilitation of the greenback, in the form of either weekly visits to the unemployment office or reduced purchasing power.

Another topic not elaborated in this chapter is the effect of flexible exchange rates on inflation, but only because I have thoroughly reviewed this matter in my book *The International Economy: A Modern Approach*. A short review of the fundamentals might be useful, however. The general price level of a country is some kind of weighted average of all prices, domestic as well as foreign. Thus when the currency of a country depreciates and the price of foreign goods in terms of domestic currency increases, the general price level automatically increases, if *everything else remains the same,* a more than reasonable assumption for the short run. What happens when the currency of a country appreciates? In this case, does the price level decrease? In the United States over the past six or seven weeks (from about February 1, 1980) the dollar has been slowly appreciating with respect to many other currencies and to gold; at the same time, the price level has risen faster than at any time in the past fifty years. But this, of course, should have been expected. Under most normal and abnormal circumstances in the real world, prices rise; they do *not* fall! Effectively they are on a ratchet, and it would be apparent to anyone bothering to notice that the passage from fixed to flexible exchange rates came just in time to palpably boost the inflationary pressures in the world economy.

It is also important to make clear that when countries with even large unemployment resort to expansionary economic policies, often their currency depreciates, raising the price of imports and imparting the mechanical boost to the price level outlined above, as well as severely complicating the wage bargaining process. As a result, governments find that the freedom which a flexible exchange rate was supposed to provide actually ties their hands in the long run as far as demand management is concerned.

Appendix 6A: The Real and Money Rates of Interest

Let the money price of a *physical* asset delivered at time t be $p_1(t)$. Let us also consider another asset, money, which is taken as numeraire, with a price of unity. Let $r_1(t)$ be the rental rate for the services of the physical asset during period t, and let $r_2(t)$ be the money rate of interest, the rental for one unit of money for one period. It should be emphasized that these are money rentals and net of depreciation. Now observe that the amount of the asset in period $t + 1$ is $A(t + 1) = A(t) + \Delta A(t)$. It should be obvious that $\Delta A(t)$ is

$$\Delta A(t) = \frac{r_1(t)A(t)}{p_1(t)} \tag{6A.1}$$

This is so because $r_1(t)A(t)$ is the total money rental on the physical asset during the period lasting from t to $t + 1$, while the rental in terms of goods, or the real rental, is given in equation 6A.1. Thus for $A(t + 1)$ we have

$$A(t+1) = A(t) \left[1 + \frac{r_1(t)}{p_1(t)}\right] \tag{6A.2}$$

In this expression $r_1(t)/p_1(t)$ is analogous to an interest rate: it is known as the real rate of interest.

Next observe that 1 monetary unit buys $1/p_1(t)$ units of the physical asset A. During the period from t to $t + 1$ this yields money rentals of $r_1(t)/p_1(t)$, and at the end of the period the asset can be sold for

$$p_1(t + 1) \frac{1}{p_1(t)} = \frac{p_1(t + 1)}{p_1(t)}$$

The total money yield on this asset is then

$$\frac{r_1(t)}{p_1(t)} + \frac{p_1(t + 1)}{p_1(t)}$$

In equilibrium this must be equal to the yield on any asset, including a financial asset. Thus we have, with p_2 the "price" of money:

$$\frac{r_1(t)}{p_1(t)} + \frac{p_1(t+1)}{p_1(t)} = \frac{r_2(t)}{p_2(t)} + \frac{p_2(t+1)}{p_2(t)} = r_2(t) + 1 \tag{6A.3}$$

The right-hand side of equation 6A.3 results from the fact that $p_2(t) = p_2(t + 1) = 1$. From this expression and $\Delta p = p(t+1) - p(t)$, we get:

$$\frac{r_1(t)}{p_1(t)} + \frac{\Delta p_1(t)}{p_1(t)} = r_2 \tag{6A.4}$$

where $\Delta p_1(t)/p_1(t)$ is the rate of increase in the price while $r_1(t)/p_1(t)$ is the real rate of interest. As specified earlier, $r_2(r)$ is a money rate of interest. Now we see immediately from the above that if initially the real and money rates of interest are equal and $p_1(t)$ begins to increase, then the money rate will become larger than the real rate. The reason is simple. If we take $r_2(t)$ constant to begin with and $p_1(t)$ begins to grow, then the left-hand side of the above relationship is increasing, which means that the yield of the physical asset becomes greater than the yield of the financial asset. Under the circumstances, it pays to borrow money to buy the physical asset, and this in turn drives up the rate of interest $r_2(t)$.

We can approach this matter another way. The present value $V(0)$ of a future unit of any good is the amount of money that must be set aside today in order to yield an amount of money equal to its price at that date, or

$$V(0) = p(t) \exp\left[\int_t^0 r(\Theta)\, d\Theta\right]$$

or

$$\ln V(0) = \ln p(t) + \int_t^0 r(\Theta)\, d\Theta$$

Differentiating, we get

$$\frac{1}{V(0)} \frac{dV(0)}{dt} = \frac{1}{p(t)} \frac{dp(t)}{dt} - r(t)$$

In line with Keynes, Samuelson (1937) has defined the commodity rate of exchange as

$$r^* = -\frac{1}{V(0)} \left(\frac{dV(0)}{dt}\right)$$

Thus we get

$$r = r^* + \frac{1}{p} \frac{dp}{dt}$$

If we interpret the commodity rate of exchange as the real rate of interest, this last expression is the continuous form of the result given in equation 6A.4.

7 The Oil Companies, OPEC, and the Rotterdam Spot Market

The achievements of OPEC are no longer a matter for debate. An organization which was founded to protest the loss by several producing countries of a few cents of revenue per barrel because of the pricing policies of some multinational oil companies has managed to achieve an extraordinary increase in the earnings of its members without the permission or good wishes of either the oil companies or the industrial world. Moreover, instead of contenting itself to be just another pretentious talk shop, this international organization features as much muscle as mouth. An OPEC meeting is always the time for the decisionmakers in the industrial world to hold their breath and on more than one occasion to gnash their teeth, as they helplessly await the news that the forthcoming oil price will not be within the range that they consider acceptable.

The history of the oil producers' organization dates from 1960, and although it sometimes appears that OPEC was quiescent until 1973, now it seems clear that the thirteen years between these dates were not entirely wasted. Among other things, many of the gentlemen now deciding among themselves just what the world's motorists should pay for their gasoline were educating themselves in the complexities of the international oil market, and educating themselves, it turned out, better than their counterparts in the industrial countries. By way of coincidence, some of the so-called leading oil experts of the Western world were busy during those years establishing the reputations that many of the governments of the industrial world made the mistake of relying on in that traumatic autumn of 1973, when the industrial world entered the first phase of a 400% oil price increase. For instance, in the late 1960s M.A. Adelman of the Massachusetts Institute of Technology was feverishly assuring one and all that the price of oil would sink to $1/bbl by the early 1970s. His reasoning was as follows. The average production cost of Middle East oil was somewhere in the range of $.08/bbl to $.10/bbl, as opposed to $.50/bbl in Venezuela and $3/bbl in the United States. The posted or listed price of marker crude was $1.80/bbl, but a great deal of free-market oil was being sold for a price of about $1.40/bbl because of a tendency toward surplus on the world markets at that time. Thus, on the basis of his estimate of the potential supplies that could be brought to market, in addition to what he claimed was the patent impossibility of forming a workable cartel, Adelman's conclusion could only be that the force of competition would eliminate the excess profits implicit in the prices cited. On average, these profits, expressed in rates of return, came to 70 percent in the Middle East, 20 percent in Venezuela, and 10 percent in the United States.

Whether the oil-producing countries accepted this analysis or not is irrelevant, because they were already thinking in terms of much larger revenues than they were getting. Early OPEC experiments with prorationing failed to produce the desired results, but world economic conditions, especially if viewed over time, were working in favor of these countries. To begin, the demand for oil was growing rapidly; between 1969 and 1973 it reached 10.6 percent per annum, which reflected the accelerating economic growth in Western Europe. More important, oil production in the United States peaked in about 1970, which left that country with no spare capacity to cushion a possible downward adjustment in Middle Eastern production. Similarly, Russia's internal demand for oil grew to such an extent that it became necessary for the USSR to stabilize its exports of oil to noncommunist countries. Because of the changing geographical disposition of Russian oil, which increasingly tended to feature production in remote areas (for example, western Siberia) where the costs of both drilling and transport are high, often some of the southern border areas of the USSR imported oil from Iraq or Iran. This was an important deviation from the situation five years or so earlier, when Russian crude competed against that of the Middle East on world markets.

Then, too, the theoretical conditions which should apply in order to make a cartel effective were coming into existence, in particular a cartel composed of producing countries as opposed to private businesses. By 1973 OPEC members controlled 70 percent of world reserves, 54 percent of production, and 84 percent of exports. Equally important, by assuming responsibility for all the oil within their national boundaries, they effectively reduced competition in the world oil market by squeezing out those independent oil companies whose general policy was to produce as much oil as possible and who were causing the major oil-producing companies a great deal of annoyance. Moreover, as explained earlier, the elasticity of demand for crude petroleum was extremely low. If we define elasticity as the percentage change in demand caused by a 1 percent change in price (where it is understood that this elasticity is negative), then the *short-run* elasticity of demand for oil was estimated to be between -0.10 and -0.15, while the long-run elasticity was somewhere on the order of -0.3 to -0.4. Consider what this means: If the price of oil were raised by 1 percent, even in the long-run, demand would fall by only 0.3 to 0.4 percent! Accordingly, revenue would rise since the percentage fall in demand was less than the percentage rise in price.

The Iranian revolution and the sudden fall in the supply of Iranian oil also provided an insight into the magnitude of the very short-run elasticity (that is, the *impact* elasticity) of this commodity: A shortfall of 2 percent in the world supply of oil caused a price rise of 25 percent which, when taken at face value, indicates a smaller short-run elasticity than that given above. One key reason is the presence in the industrial world of a great deal of expensive, long-life equipment that had been painstakingly designed to rely on what appeared to be an infinite supply of low-priced oil.

The price elasticity of the supply of crude oil is also very low. Certainly the price rises of 1979 and the first part of 1980 have not caused large amounts of new supplies to be brought to the market. The problem here is simply that there are no new supplies comparable to those which were waiting to be discovered a few decades ago. Moreover, for the major oil users—Japan, Western Europe, and the United States—there is no point in worrying about the lack of prospecting that characterizes the world oil economy today since, even if oil were found, most likely it would be in "politically unreliable" localities. Here it should be noted that the term *politically unreliable* can be toned down slightly and applied to such domains as the western provinces of Canada (and perhaps also the maritime provinces, should the offshore prospects there live up to rumors), since these provinces seem to have adopted a "charity begins at home" attitude as far as oil is concerned.

It should also be apparent that the price elasticity of supply of substitutes for oil is extremely low. Both the United States and Sweden are energy-rich countries. Sweden, for example, has not only Europe's largest supply of uranium (which presently it has no intention of exploiting) but also huge supplies of peat bog, unexploited waterpower, the area and capacity for growing energy forests, high winds suitable for the operation of windmills and the like, and the technology and organizational ability to take advantage of all these *and* to greatly increase the already high degree of energy conservation. But despite the fact that six years have passed since the first major oil-price rises, only a trivial fraction of this potential has been realized. In fact, if the oil price continues to rise at its present rate, it is conceivable that such a large amount of Swedish resources in the form of production capacity will have to be utilized to produce the products to pay for oil imports that the country will not be able to afford to make the necessary investments in alternative energy sources unless (which is quite likely) the burden associated with these investments is passed to future generations in the form of having to amortize and make the interest payments on a staggering foreign debt. This situation is also faced by many LDCs; however, they will not be able to make these investments in any case, since for demographic and other reasons their capacity to repay foreign debt will probably be much worse in the future than at present. Still, like everyone else they will need oil; and according to a recent estimate by Electricité de France, third-world oil demand in 2010 could exceed today's total world demand.

The Majors

Even a cursory analysis of OPEC must necessarily build on the activities of a group of oil companies known as the "majors" or "seven sisters." I say *necessarily* because it was the scaffolding created by these giant firms that was taken over and utilized by the oil-producing countries. On the date that OPEC officially came into being, the majors owned 84 percent of world crude-petroleum production facilities outside the United States and communist-bloc countries and possessed or controlled 74 percent of refining and 70 percent of marketing capacity. To use Biblab Dasgupta's superb terminology, " . . . they knew everything about oil and did everything there was to do to run the industry."

Five of these firms were (and are) based in the United States: Standard Oil of New Jersey (sometimes called Exxon, Esso, or Jersey), Socony Mobil Oil Company, Standard Oil of California (sometimes called Socal or Chevron), Gulf Oil, and Texaco. One is British (British Petroleum), and one is Anglo-Dutch (Royal Dutch Shell). It is possible to argue that another company belongs to this select group, the Compagnie Française de Pétroles (CFP), of which 35 percent is owned by the French government. To concretize Dasgupta's observation, these firms explore, produce, transport, refine, and market. They deal in everything from the crude material to petrochemicals; at present they are moving into other energy fields as well as industries that have little or nothing to do with energy (such as publishing). They also do enormous amounts of research and are the repository of the technology and financial resources required to make the costly high-risk investments that characterize this industry and its affiliates. In a broader sense they can be thought of as the mobilizers and deployers of the technical and organizational skill on which the world oil industry has been built, although the world is full of people who refuse to acknowledge this fact. A few significant statistics about the U.S. majors are given in table 7-1.

Since the majors are so important, a slightly closer look at them seems warranted. To begin, we consider Exxon. This company dates from 1911 and the breakup of Standard Oil of New Jersey by the U.S. Supreme Court.

Table 7-1
Some Figures concerning the Major U.S. Oil Companies, 1978 and 1979

Year	*Data*	*Exxon*	*Mobil*	*Texaco*	*SoCal*[a]	*Gulf*
1978	Sales (000)[b]	60,334,527	34,736,045	28,607,521	23,232,413	18,069,000
	Assets (000)	41,530,804	22,611,479	20,249,143	16,761,021	15,036,000
	Net Income (000)	2,763,000	1,125,638	852,461	1,105,881	791,000
	Employees	130,000	207,700	67,841	37,575	58,300
1979	Sales	84,350,000	47,900,000	39,096,000	31,800,000	26,137,000
	Net Income (000)	4,295,000	2,010,000	1,759,000	1,785,000	1,322,000

[a]SoCal is the usual designation for Standard Oil of California.
[b]Sales, assets, and net income figures are in current dollars.

Exxon is active over the entire world, but its most important assets are, or were, in the United States, Venezuela (until 1975), Saudi Arabia (where it has a 30 percent share of Aramco, as is commented on later), Iran, Abu Dhabi, Libya, and Iraq. Of course, progressive nationalization has captured some of these assets, but Exxon is still one of the most powerful firms in the world and, in light of the importance of energy, its growth, sales, and profit rate, it belongs at the top of the industrial league.

Thanks to its North Sea properties, Shell may now be larger than British Petroleum (BP). Initially Shell's principal assets were on Sumatra in Indonesia (or the Dutch East Indies, as it was known at the time). Then, through the French branch of the Rothschild banking family, Shell acquired a large interest in Russian oil. Later Shell acquired considerable properties in Venezuela and Mexico. Today the most important properties of the firm are in the Middle East, the United States, and Nigeria. Shell was part of the consortium operating the Iranian oil fields and had a large interest in the Iraq Petroluem Company until this organization was nationalized in 1972. Shell is a very flexible firm and tries to produce as many of the most profitable light products as it can. Similarly, BP has important assets in Africa and was the most important member of the Iranian consortium. This company always has had close ties with the British government, and at one time the British admiralty bought into the company for the purpose of securing supplies of oil, particularly for British vessels operating in the Middle East. The original name of the company was Anglo-Persian, and it has been very active in Kuwait and Abu Dhabi as well as Iran and as a major seller of crude to other companies. Quite naturally, it plays a very important role in the British sector of the North Sea and has an important interest in Alaskan oil.

Next we come to Texaco, whose ranking among the great oil companies is after Exxon and Shell but, at least until recently, slightly ahead of BP. Texaco started out by concentrating on the refining and distributing side of the oil business, but eventually it was able to trade some of these assets for a slice of the huge reserves that came under the control of Standard Oil of California (SoCal). SoCal has important properties in the United States and has been active in the exploitation of Indonesian oil, but Socal's greatest success was registered in Saudi Arabia where, by virtue of not being involved in the so-called red line agreement, which was designed to inhibit competition in the Middle East, it was able to make a number of extremely profitable acquisitions. Texaco and SoCal have a joint marketing and refining operation, Caltex, and recently they sold their Spanish refinery to the government of that country and their 19 percent share in French refining and marketing to France's Elf-Acquitaine.

The smallest of the majors are Mobile and Gulf. Mobile was a product of the breakup of the original Standard Oil Company and was without crude at the start. Eventually, however, Mobil purchased considerable assets in the United States, North Africa, and the Middle East. Gulf was founded at the same time as Texaco, and has been associated with the well-known Mellon family of Pittsburgh Pennsylvania, whose other interests include Bethlehem Steel and the Aluminum Corporation of America. For a long time Gulf's most important interests were located in Kuwait where, despite the fact that its properties have been nationalized, Gulf retains some useful connections. This company has also been very active in Angola, but is has been said that on the whole this firm is in the process of abandoning its foreign commitments. Gulf has sold its refining and marketing outlets in France and Spain and is trying to sell a portion of its Korean assets. Gulf has only two major refineries left in Europe, in the United Kingdom (where its North Sea oil is processed) and Denmark.

Although these companies are independent in both theory and law, it is impossible to avoid noticing that the Rockefeller family is or has been heavily involved with at least four of them; the firm of Price Waterhouse & Co. has done a good deal of the accounting for the majors; and banks such as Chase Manhatten, First National City, Morgan Guaranty, Chemical Bank, and Continental National Bank and Trust of Illinois are each involved with at least two of the majors, as well as some of the larger of the minors. Moreover, these companies were closely associated in the most important consortia of the 1950s and 1960s—the golden age of oil. Apparently a number of people have been disturbed by this cross fertilization, and this includes not only Mr. and Ms. Consumer and the antitrust departments of various governments, but also more than a few of the directors of these companies whose competitive instincts have been held in check by the need to occasionally cooperate with their rivals.

It could be argued, though, that in certain enterprises too much competition is not conducive to the good health of the majority. Take a situation where infantry platoons are moving successively across a firing range on which a number of targets have to be engaged and neutralized. The platoon carrying out this assignment in the shortest time will be rewarded by a warm smile from the base commander. The purpose of competition here is to increase the performance of the platoons, their endurance, target selection, firing accuracy, and so on. But this kind of competition, and the attitudes which it engenders, can be counterproductive when these platoons are operating side by side in a combat situation, because rapid movement by one unit relative to another might increase the exposure of all the enemy fire. Here coordination becomes the key to successful performance. The same may be true, on occasion, in the oil industry, where enormous successes in prospecting or technological breakthroughs in the building of mastodon refineries by a large company with access to plentiful financial resources might be disastrous for everyone, including that company.

Because of their importance in one respect or another, a few other organizations should be mentioned here. Among private firms, Continental Oil (Conoco) and Atlantic Richfield have received considerable attention in the business press. Conoco is a former distribution company in the Standard Oil group that has had important interests in Libya, Qatar, and the Norwegian sector of the North Sea. Atlantic Richfield has been heavily involved in the United States, Iran, Indonesia, and Venezuela, and it was the first company to find oil in Alaska. The chairman of Atlantic Richfield, Thornton Bradshaw, is in some circles regarded as the most innovative thinker in the U.S. business world.

Government organizations deserve a place in this summary, since it is becoming apparent that some oil-producing countries, through *their* state-owned oil companies, prefer dealing with state-owned oil companies in the oil-importing countries. Some of the most important of these companies on the producer side are Nioc in Iran, founded in 1954; and CVP in Venezuela and KNPC in Kuwait, both founded in 1960, the same year that OPEC came into being. Petromin in Saudi Arabia, probably the most important of these organizations, started up in 1962. Sonatrach started in Algeria in 1963; Inoc, in Iraq in 1965; and Linoci, in Libya in 1969. The most important government companies in the oil-importing world are Elf in France and Eni in Italy. The latter is of particular interest since, under the leadership of Enrico Mattei, this company attempted to bypass the majors and deal directly with the oil producers at a time when sensible people found it in their interests to avoid even thinking about such behavior. As it turned out, however, Mattei overextended himself on a number of projects and died under circumstances that were, if not mysterious, at least melodramatic. Another interesting firm is the French CFP, which is partially owned by the French government and occasionally referred to as the "eighth sister." This company, like Elf, is working to give France an independent oil policy, specifically one that is independent of the United States and that will ensure deliveries of oil to France in case "radical" governments or parties become more important in the countries around the Persian Gulf.

Just as there is a great deal of money to be made by owning oil-producing or-processing assets, trading in oil shares has proved one of the most profitable speculative activities of the past few years. In the first four months of 1980 the Financial Times' actuaries oil index increased by almost 20 percent while the all-share index increased by 13 percent. Looked at over a slightly longer time horizon, this index has increased by 300 percent since 1973, while the all-share index has increased by 50 percent. There are few, if any, industries with this record of achievement on the United Kingdom stock markets, and a similar phenomenon can be observed on the stock markets of the United States and elsewhere.

Apparently any rumor of a cutback in OPEC oil production will send a flood of speculative cash into the markets, and the special items of interest are those oil companies whose physical assets are to be found in "safe" locations (that is, out of range of direct OPEC administrative influence) and those who are in line for a takeover bid by prestigious industrial or financial organizations. The North Sea has become a happy hunting ground for this type of activity, and some of the stories coming out of that region take on a peculiar similarity to those that were circulating in Geneva during the high tide of Bernie Cornfeld's venture into "people's capitalism." Just recently a small exploration and production company made a major share offering that was oversubscribed forty times, although that company has yet to find oil in the North Sea or anywhere else.

For a through-the-keyhole examination of the petroleum industry, see Sampson's interesting and well-written book (1975), where economics is put in a form that might someday be suitable for a prime-time TV drama. In a certain sense, this approach is useful, because in the world of giant oligopolistic oil companies with their market sharing in an international setting, double crosses, price wars, misunderstandings, triumphs, and multimillion-dollar blunders, the foibles and predilections of a relatively few individuals seem to play an important part in explaining why certain things happened when and how they did and why equally likely events did not take place. At the same time, though, it should be stressed that the oil market has its own interior logic which derives from the tremendous pace of industrial development in North America, Europe, and Japan in the years following World War II. The relatively inexpensive oil that fueled this development *was* going to be found and processed regardless of who occupied the executive suites at Gulf or Texaco and regardless of their personality splendors or defects. Although the special conditions existing in the Southwest of the United States or the arctic might occasionally permit a semblance of tolerance for independent thinking, the collective leadership that is becoming the fashion in all facets of economic, social, and intellectual life should serve to keep all except conventional team players out of the board rooms of the larger oil companies in the near future.

An Historical Excursus

I now propose to back up slightly and take an abbreviated look at some important historical developments in the world oil market, beginning at the point where oil was first discovered in commercially attractive quantities. At the beginning of the twentieth century about one-half of the world's oil production was accounted for by the United States and Russia. The U.S. oil industry was dominated by Standard Oil, and the key name behind that enterprise was Rockefeller (in this case, John D.). In 1865 a refinery had been built with the company name *Rockefeller and Andrews;* by 1878 this organization, which is now called Standard Oil, controlled 98 pecent of all the refineries and oil transport facilities in the United States. But in 1911 Standard Oil's financial power and the aggressive practices of its management caused the U.S. Supreme Court to intervene and break the company into three parts: Standard Oil of New Jersey (Exxon), Standard Oil of New York (Mobil Oil Corporation), and Standard Oil of California (Chevron). As is to be expected, however, these firms occasionally found it useful to assist one another; and the Rockefeller family still has a modest interest in their affairs.

Outside the United States the most important oil producers were Royal Dutch, and Shell. The latter, a British firm, merged with Royal Dutch in 1907 to form Royal Dutch Shell, with Dutch interests holding 60 percent of the equity of the new company. The feedstock base for these companies was largely in Indonesia (at that time the Dutch East Indies), but the resources of that region became secondary after the first major discoveries in the Middle East in 1908. These were in Iran, and in order to facilitate their exploitation, the Anglo-Persian oil company was formed. The peace settlement following World War I gave Britain and France a large part of the Middle East, but the U.S. oil companies, glimpsing the prospects of the region, eventually managed to get the U.S. government to back an open-door policy which would allow all companies, large and small, rich and poor, from any part of the world, a fair chance to compete for oil belonging to Arabs and Persians. As far as is known, however, this generosity did not extend to the indigenous peoples of these territories.

Eventually the Iraq Petroleum Company (IPC) was formed, which was the first of the important joint ventures by major international oil companies. It was important in the sense that it showed the ability of the oil men to recognize the economic (and perhaps political) strength of each other, and to accept that cooperation, where possible, was often more conducive to the profit-making process than conflict. To begin, equity in the IPC was divided as follows: British Petroluem, 23.75 percent; Shell, 23.75 percent; Compagnie Française de Pétroles, 23.75 percent; Exxon, 11.875 percent; Mobil 11.875 percent; and the perennial 5 percenter, Mr. Gulbenkian, took his 5 percent. As for the "open door," that was slammed and bolted, and a no-welcome sign was hung on the outside. Instead the red-line agreement of 1928 was concluded in which these companies promised not to compete with one another over concessions within an area comprising what at one time had been the Turkish Empire, and which included Iraq, Saudi Arabia, and Bahrain, but not Kuwait or Iran.

What the red-line agreement did not do was to shut out firms not included in the joint venture, assuming that they were strong enough to push their way in. One company in this category was Standard Oil of California, which already had huge assets in east Texas, as well as access to considerable financial resources. This firm gained major concessions in Saudi Arabia, but lacked the means to distribute and market the oil that would soon begin to flow from its new properties. This dilemma was relieved to some extent by its selling half its concession to Texaco, which was also outside the red-line agreement, in return for access to some of Texaco's marketing outlets. This was a superb example of exchanging future revenues for present revenues on the part of Standard's leadership, and present revenues for future revenues from Texaco's side; but eventually, given the huge amount of oil in Saudi Arabia, the desire of other companies to share in this bonanza became so overwhelming that in the interests of maintaining harmony in the oil-producing community a new consortium was formed. This was the famous Aramco, which contained Mobil and Exxon in addition to Standard of California and Texaco.

The entrance into Aramco was a decisive chapter in the history of world oil because of the importance of Saudi Arabia as a producing country and because the ties established between the oil companies in that country affected their dealings elsewhere in the world. Another milestone of this type was the "Achnacarry" or "as is" agreement. This covenant, which was the brainchild of Sir Henri Deterding, was entered into at Achnacarry Castle in Scotland. In his youth Deterding had been a workaholic, but later in his life he became a playboy; and, in all likelihood, one purpose of this agreement was to provide him with more time for his private life. (Still later Deterding took to flying into rages when he thought of how little people wanted to work, and after falling in love with Nazi Germany he abandoned his British hideaways for Continental luxury.) In any event, Deterding's company—Royal Dutch Shell—agreed to avoid territorial infringements where the marketing prerogatives of Exxon and BP were concerned, and they reciprocated. Market shares were frozen, and the three companies agreed that the most worrisome problem, which was overcapacity in both refining and distribution, was to be attacked by removing a large part of the uncertainty associated with future prices. Where possible, these prices were to be stabilized. In addition, existing facilities would be made available to competitors (that is, "friendly" competitors) on a favorable basis, but not at less cost than the actual cost to the owners of these facilities.

The Achnacarry agreement was about principles and, in addition to the above points, specified territorial rights and responsibilities. This agreement was followed by a number of "instructional agreements" which specified the basis on which quotas were to be drawn up, how the quotas were to be extended (at the expense of the firms outside these arrangements), cross payments in the event of over- and underproduction, and so on. Here it should be noted that although only Shell, BP, and Exxon had been privy to the original agreements in 1928, by 1932 all except one of the seven sisters had joined. Actually they could hardly avoid having something to do with the agreement, since its founders made it quite clear that, whenever possible, uncontrolled outlets and assets should be transformed into controlled ones without delay; where outsiders had crude, they were, when possible, to be denied refining facilities; and if they had refineries, they were to be denied crude. This particular approach to the oil business was mostly intended to be applied with smaller companies and wildcatters, but if any of the larger companies wanted to defy the others, it was made to understand that it would be given a taste of the same medicine.

All this might sound a little nasty, but it could be argued that in certain circumstances too much competition worked to the disfavor of the consumer. True, the presence of large numbers of independents scouring the face of the globe searching for oil, finding it, and then doing everything humanly possible to have it flood the world markets favors the consumer, but it may not favor the consumer in the long run. If there are too many producers and the price of oil is pressed down too low, companies find themselves with too much capacity and/or too much unprofitable capacity, and so they stop investing. Eventually this could lead to sharp price rises because of supply deficiencies and serious inconveniences for those people who in some way are dependent on cheap oil. Drastic price falls could also lead to oil fields being abandoned before the last drop of exploitable oil had been squeezed out of them: if the lifting and transport equipment and infrastructure in a field are removed or allowed to fall into dilapidation, and then later, even if the price of oil rises, it may not pay to make the investments necessary to retrieve the remaining oil.

Although we cannot settle here whether the consumer has been aided or cheated by market and ownership practices in the oil industry, history seems to indicate that the larger oil companies were adequately served. Even though the Depression caused the world demand for oil to falter and there was some overcapacity in crude-oil production in Texas, the Middle East, and Venezuela and, in the eyes of the majors, too much government-inspired refinery construction in France and Italy, it was possible to hold the price of oil well above Middle Eastern production costs, particularly in the last part of the 1930s. This was quite a feat because, relative to consumption, there was a huge supply of cheap, easily exploitable oil in the world at that time. Finally, in the period just prior to World War II the trade in oil was dominated by finished products: 48 Mtons of crude oil and 66 Mtons of oil products were traded in 1938, with the Dutch *West* Indies playing a major role on the refining side.

During World War II, world oil production increased until by 1945 it was almost 100 Mtons/yr larger than in 1939. Most of this increase took place in the United States, Venezuela, and Iran. The real escalation in civilian consumption was yet to take place, however; and for both the United States and Europe it began with the war in Korea—an event which assured many people that the noncommunist world could not afford any serious economic downturns (of the type that constantly threatens to take place nowadays) and that rapid economic growth was in the cards for Western Europe.

But before this upswing took place, Venezuela, the second largest producing country in the last part of the 1940s, introduced a 50 percent tax on oil profits, basing this tax on "Caribbean" prices. The oil companies did not take kindly to this imposition, but in the immediate postwar period, when talk of fraternity and human rights filled the air and the airwaves, the governments of the United States and Britain were hardly prepared to tolerate an ugly scene between the big oil firms and the friendly government of a country whose people viewed the rapid depletion of their most valuable resource with mixed feelings at best. Several years later the Saudi Arabian government demanded a larger share of Aramco's income and compelled the consortium to introduce a posted price for oil. This price was intended as a reference price on which to levy a 50 percent profits tax, to replace the previously imposed royalty as the main levy on Aramco.

Although the companies resented the independence of the Saudi Arabian government, they did not contest the tax because U.S. tax laws permitted taxes paid abroad to be deducted at home; in particular, this meant that integrated companies could shift a part of their profits from processing to that part of the firm producing crude where, for tax purposes, these profits would be decreased by the provisions for *depletion* (which functioned so as to compensate a firm for using up its wasting or depletable assets in the same way that provisions for depreciation compensate a firm for using up its fixed assets). The oil companies handled this by making sure that the posted price for the crude they sold to their refineries (which might be called a transfer price, since this transaction is interior to the firm) was considerably larger than the production costs for this crude. Thus the costs for their refineries were inflated, which reduced profits and thus taxes; while the correspondingly larger profits that accrued to the production of crude could be offset, or partially offset, by depletion allowances. This accounting technique was introduced in almost all the main producing countries with the exception of Iran.

In Iran arguments over the division of oil revenues led to nationalization of BP's properties in 1951. The Iranian government did not bother to offer BP (or Anglo-Iranian, as it was formally labeled) compensation, reasoning that the oil in that country belonged to the Iranian people in the first place. Despite the inconvenience felt by the oil-producing fraternity, they did not make a spectacle of themselves by demanding restitution; nor was it necessary. In the two years that followed the nationalization, Iranian revenues were less than a single day's royalties under the old system. What happened was that the production of crude oil was increased in other oil-producing countries, particularly the United States, and after a while world production was back up to a level where virtually all categories of consumer were satisfied. Under the circumstances, neither BP nor the other operating companies expended a great deal of effort trying to negotiate a settlement; they simple sat back to await the inevitable.

The inevitable arrived in 1954 when the Iranian government was overturned in a coup organized by one of the era's most publicized intelligence organizations. A Western consortium then took over operation of the country's oil-producing assets, with British Petroleum, which previously had 100 percent of the concession, becoming principal concessionaire with 40 percent. The other owners were Shell (14 percent), Gulf (7 percent), Mobil (7 percent), Exxon (7 percent), SoCal (7 percent), Texaco (7 percent), CFP (6 percent), and Iricon (5 percent). This organization continued to function until a few years ago, although after 1960 its principal assignment was to take and carry out orders from the Iranian authorities. One reason is that formally the nationalization of both the oil fields and the refineries stayed on the books, and eventually the day arrived when both sides saw that the master-servant relationship established in 1954 had been nullified. All that came later, however, and in 1954 other oil-producing countries, in observing the details of the drama in Iran, called off whatever nationalization plans they might have been making and settled for the Venezuelan-Saudi Arabian formula of a 50-50 profit split.

The key element in upsetting the nationalization attempt of the Iranians was the ability of the larger oil companies to mobilize production elsewhere in the world, from either their own installations or those of independent producers, and not the interference of gunboats or marines (although there was some talk of eventually resorting to these). But given an exponential increase in the world consumption of oil of about 8 percent a year, with a large part of this increase scheduled to come from the huge reserves of the Middle East, it must have been elementary for anyone giving the matter a few seconds' thought to realize that in a decade or so world demand for oil would be so large that in the event of a nationalization attempt by one or more of the leading Middle East producers, it would be quite futile to invoke the practice of increasing production elsewhere. This is so for two reasons: (1) For all practical purposes, there was not enough oil elsewhere to compensate for a fall in Middle East production except possibly in the United States and (2) not only would U.S. production have to be raised, but also U.S. consumption would have to be lowered in order to provide for the needs of consumers elsewhere. Given the growing cost of U.S. oil and the psychology of the U.S. automobile owner, it was very dubious whether this program could be carried out. Thus, as early as 1954 the handwriting was on the wall; but, for reasons that are not entirely clear to this day, both the directors of the major oil companies and leading civil servants in the main industrial countries chose to ignore it.

Another factor of great importance for the present exposition is the decrease in U.S. imports from the Middle East in the beginning of the 1950s resulting from a misguided belief by certain parties in the United States in self-sufficiency in oil—a belief based on another equally absurd belief that there were infinite supplies of oil in North America. In 1959 an import quota was placed on oil coming into the United States. The most widely published arguments in favor of this quota were tied to defense issues, in particular having enough capacity for producing crude that the country would be independent of foreign imports in time of war. Ostensibly quotas would lead to higher oil prices, which would cover the higher cost of drilling deeper wells and offshore exploration. Thus, should war come, more reserves would have been located. But the major oil companies were also heavily involved in this matter for economic reasons. The apparatus they had built up to control oil prices in the United States could be threatened at any time by heavy imports of oil into the United States. And in light of their own revenue requirements, had free access to the U.S. market been allowed, the governments of the oil-producing countries would have insisted on taking full advantage of this opportunity. It hardly needs to be pointed out that the long-run interests of the United States would have been far better served if, instead of a quota on imports, the production of crude petroleum in the United States had been cut during the 1960s. This would have stretched out the limited supplies of the United States and probably delayed the peaking of domestic production. In addition, it would have retarded the growth in OPEC's militance, since OPEC's revenues would have been much larger.

Another result of the U.S. import quotas was that many "independent" U.S. companies were able to establish themselves in the Middle East and North Africa. The governments of some of the oil-producing countries felt that the majors had been sympathetic to the quotas, and the independents presented themselves as the best device for breaking into the European market. Whether this was true or not was uncertain, since the French and Italian state-owned companies were *also* active in the area; but there certainly seems to be some truth in the claim that the undermining of the control of the majors over a large part of the oil supply of the world began with the import quotas introduced by the United States. In connection with the mustering of the independents and their designs on the European market, something about the changes in the structure of the world oil trade should be mentioned. In particular, the oil being consumed in Europe was being refined to a progressively larger degree in Europe. Behind this phenomenon was the rapid growth in demand for different types of oil products, particularly gasoline, which could be produced more economically in Europe than in an oil-producing country (because of certain advantages of large-scale production) and the development of huge oil tankers, which reduced the cost of transporting crude relative to oil products. Eventually imported refined products fell from being 45 percent of total consumption in 1937 to 25 percent in 1953 to near self-sufficiency by the end of the 1950s. The importance of this development can be realized when we note that in 1962 in Europe, twenty times as much refined products were being produced as in 1946.

The Arrival of OPEC

We open our discussion of OPEC by saying something about the price of oil Figure 2-1 illustrates the development of this price over time, but it gives no indication that, along with the 1973-1974 oil-price rises, the most important Post-World War II price event was the setting of the money price of oil at $1.80/bbl in 1960. (Here it should be made clear that $1.80/bbl is the price of Arabian Light oil, and this serves as a reference price for other oils. In this sense it is sometimes called a *marker crude*.) Although not indicated in the figure, setting the price at $1.80/bbl involved a 15 percent decrease from the price prevailing at the beginning of 1959; and if we consider the steady rate of increase of the price level in the industrial countries, a constant money price meant a falling real price or, as it was sometimes expressed, a fall in the terms of trade or in the purchasing power of oil.

There were several reasons for this reduction in the oil price. The first that can be cited is the oversupply of oil generated by the independents, some of whom were protected by the oil-producing countries from the wrath of the majors. But, in hindsight, more important were the phenomenon of accelerating industrial growth in Europe and Japan and the intention of the oil companies to make sure that the industrial structure in these places was based on oil, not coal, nuclear power, or some other energy source. Another interesting facet of this topic is that in early 1959, when the posted price of oil was $2.08/bbl, a great deal of oil was being sold at the free-market price, which averaged about $1.30/bbl. Since oil firms were paying to the producer countries taxes based on the posted price, the share of profits going to the producer countries was considerably higher than 50 percent. The oil companies corrected this situation by decreasing the posted price, with Exxon taking the lead. Some economists maintain that this price decrease, which was motivated by the desire of the oil firms to obtain a few cents more in profit per barrel, was the specific cause of the formation of OPEC. Certainly the foremost goal of this organization, in the first months of its existence, was to get the price of oil back up the 1958-59 level.

During the period when the oil price was $1.80/bbl, there were two extremely important OPEC conferences. At the first in 1962 (which was actually the fourth OPEC conference) it was decided that the royalty on oil production, which at that time was subsumed by the tax on profits, henceforth should be regarded as a cost of production. This decision increased the producing country's share of oil profits. At its sixth conference in 1968, OPEC produced an "Oil Policy Statement" which was no less than a declaration of intent to assume full control of member countries' own oil resources as soon as possible. If the oil companies found this intention shocking, they kept their feelings to themselves. In 1968 there was a great deal of excitement in the world, particularly on the streets of Paris and Chicago and in Vietnam. Undoubtedly there was a feeling among certain people that if the war were won in a sufficiently spectacular manner (which seemed possible at the time), it would have the same depressant effect on OPEC as the 1954 coup in Iran had had on the oil-producing countries in the rest of the world at that time.

In March 1969, Iran demanded $1 billion in royalities from the operating companies on an "anyway you can raise the money" basis, pointedly ignoring such things as the current flow of oil company revenues, profitability, and so on. Given the petroleum prices existing at that time, this injunction meant that the companies would have to lift and sell 15 percent more oil than they had planned on handling. Needless to say, this was depressing news for these companies because world oil production was still running slightly in excess of consumption at that time, and thus more money for Iran might be at the expense of stockholder earnings. Also, giving in to the Iranians would tend to place the competence of the directors of the operating companies in a bad light. Even worse, if Iran succeeded with this challenge, the other producing countries would be tempted to try the same thing.

On this latter point the operating companies were correct. Iran received $1 billion, and it was not long after this important event that the Libyan government announced a decrease in the production of oil and an increase in price, and Algeria also increased the price of its crude. On top of all this, in 1970 Iran notified the management of the consortium that its royalties would have to come to $5.9 billion over the coming five years. The oil companies' response to this shock was to notify anybody prepared to listen that this was more than the market could bear; but, as usual, they were wrong. The solution to their problem turned out to be allowing the world price of oil to increase; and while the growth in demand might have decreased slightly, the short-run elasticity of demand was such that increases in price meant more revenue—and profit—instead of less. Moreover, at that time, the independents were under such pressure by the producing countries to raise the price of oil that they could not undersell the majors. Fortunately, though, this higher price also meant that petroleum users and producers began thinking about other sources of energy, sizable increases in offshore exploration, and so forth. Otherwise the aftermath of the 1973-1974 oil-price increases might have been even more unfortunate. Another point of interest here is that any *fall* in demand which might have come about because of the increase in price was soon more than compensated for by *increases* in demand resulting from the growth in income in the main oil-consuming contries.

Next something should be said about Libya, since that country soon became the most militant of the OPEC fraternity. In 1968 Libyan production increased by 50 percent, and for a while that country became the world's largest exporter of oil. Because of the high quality of its oil and its proximity to Europe, as well as some interruptions in production or distribution elsewhere in the world (for example, the war in Nigeria and the closing of some pipelines carrying Saudi Arabian oil by Syrai), Libya's negotiating position vis-à-vis the producing companies was strengthened enormously. And when the revolutionary government came to power in September 1970, one of its first demands was that the oil firms operating in that country cut production and raise the offical price of oil.

As a result, the price of Libyan crude (which has an API of 40) was increased from $2.13/bbl to $2.53/bbl, and new taxes were imposed on oil revenues. These taxes were also made retroactive as a kind of punishment for the previous sins of the oil companies; and at the end of 1971 Colonel Gadaffi's government nationalized BP's properties. These Libyan actions are regarded by some as a turning point in the relations between oil companies and their hosts, in that these actions signified that the countries not only had the upper hand, but also knew how to use it. However, it could be argued that the Iranian action the previous year was equally decisive.

In December 1970, the OPEC meeting at Caracas, Venezuela, formally established the new relationship between companies and countries. New tax rates were established, and the companies were fenced in with some new regulations concerning how much they should charge for their products. The same theme carried over to ensuing meetings at Teheran and Tripoli in February and April 1971. One of the most notable results of these sessions was the increase in the price of marker crude (that is, Saudi Arabian Light, an API of 34, FOB Ras Tanura) from $1.80/bbl to $2.18/bbl. This was the first time that marker crude had gone above $1.80/bbl since the end of the 1950s; altogether, the average revenue of the producing countries was raised from $0.87/bbl to $1.25/bbl.

From 1971 to the Spring of 1973

From early 1971 the intention of the producing countries was to get the price of oil up to $5/bbl or $6/bbl as soon as possible. They began this putting more pressure on the oil companies to surrender their assets, indicating that these organizations would get a respite from this harassment only if they succeeded in steadily raising the price of oil. A concrete suggestion forwarded by OPEC along this line was that the firms should start thinking about transferring 20 to 25 percent of their assets in each producing country to the particular country. The only countries to be excepted were those in the process of asking for or taking more than this amount. (Libya and Algeria were already confiscating more than 50 percent of oil company properties in their countries, and as a result of the 1951 nationalization, the Iranian government already owned all oil company assets in Iran.)

Eventually the following scheme was presented to the representatives of the oil companies. Each country that had not done so would take over 25 percent of the assets of foreign companies operating on its soil. This amount would increase by increments of 5 or 6 percent a year until 1982, at which time the producing country should be in possession of 51 percent of these assets valued at market prices (and not at book value). Before we consider the other important component of OPEC's program, it can be revealed that most OPEC countries lost interest in this schedule for the acquisition of oil company resources and simply went about taking over the companies, or as much of the companies as they were interested in, when they felt like it.

The other important arrangement worked out for the oil companies by their hosts had to do with the distribution of oil between companies and the producing country. To begin, some of the oil produced in a country falls in the category of *equity oil,* or *equity crude.* This is the portion of the output "belonging" to the oil companies; generally it amounts to 30 to 50 percent of output. As shown below, the posted price largely determines the cost of oil to the firms, although these costs can be adjusted by various fees from time to time. (Algeria, for example, attached an exploration fee to its posted price at one time.) The rest of the output (total output minus equity oil) is called *participation oil,* and it "belongs" to the producing country. Of this participation oil, a portion, called *buy-back oil,* is sold by the countries to the oil firms at a price called the *buy-back price* or, sometimes, the *participation price.* This price is supposed to be negotiated between seller and buyer. In general, the oil firms are committed to buying a certain amount of oil from the countries, thus ensuring OPEC a market for a large part of its oil even during periods of oversupply. Although the oil companies have tried, on at least one occasion, to avoid buying the amount of buy-back oil that the countries insisted they buy (and which, legally, they had to buy), the general practice is to honor commitments. This is a question not merely of business ethics, but of being ensured deliveries of crude. Without these deliveries the oil firms jeopardize the profitability of their refineries.

As things now stand, the companies have a different worry—not getting enough buy-back oil. What has happened is that the producing countries prefer to sell more oil themselves: they sell on the spot market, and they sell directly to countries—from one state enterprise to another. In 1973 the private oil companies gained access to 92 percent of the total supply of world oil, whereas in 1979 this figure was 58 percent. The exact reason for this change in selling strategy is unclear, although some guesses seem in order. In 1979 Shell and Exxon alone had combined profits on the order of $10 billion, with a sizable part of this coming from their refining and distribution operations. In fact, the profits of the five U.S. majors topped their 1978 figures by a rate of 55 percent (Exxon) to 106 percent (Texaco), and in the course of this triumph Exxon became the largest industrial company in the world. The OPEC countries want some of these profits. If they can get some of these profits, perhaps they can be less aggressive in raising the oil price, which would tend to raise their popularity with their customers and thus make the governments of the oil-importing countries much more receptive to some of OPEC's political ambitions, desires, and so on in regard to such things as the Palestinian issue. Table 7-2 shows the diminishing supply of OPEC oil to the major oil firms.

One of the companies mention in table 7-2, British Petroleum, has had difficulties in Nigeria because of some allegations over its dealings with Rhodesia and South Africa. As far as I can tell, nobody was inconvenienced by the subsequent change in ownership arrangements except BP, because the oil kept flowing; BP was simply out of the Nigerian picture—at least for the time being. It is also interesting to note that some of the oil companies are helping OPEC to broaden its operations on the world oil scene by acquiring or building refineries for several OPEC countries, in particular

Table 7-2
Amount of OPEC Oil Going to the Major Oil Firms

Oil Company	*1969 (Mbbl/d)*	*1979 (Mbbl/d)*
British Petroleum	3.25	1.20
Shell	2.70	1.75
Gulf	2.00	0.80
Exxon	3.60	3.00
SoCal	1.50	2.40
Texaco	1.65	2.45
Mobil	0.98	1.40

Note: These figures are estimated from various sources. It should also be pointed out here that although some companies have raised the amount of oil they are getting from OPEC, they are still getting a smaller share of total OPEC production than they were getting previously. On the whole, in 1969 the majors handled 60 percent of noncommunist oil supplies in 1969; today they handle about 40 percent.

Saudi Arabia, both in the producing countries and elsewhere. Shell and Mobil are making preliminary deals to construct refineries in Saudi Arabia in return for guaranteed oil supplies. They are to obtain 1,000 bbl/d for every $1 million of equity investment at 1975 dollars or 500 bbl/d for every $1 million invested in current dollars, whichever is the highest. Everything considered, the only thing that prevents an even more rapid bypassing of the oil companies by OPEC is the shortage of OPEC-controlled refining capacity and/or the limited state participation in the oil sector in the importing countries. An important issue here is whether governments in Western Europe, Japan, and perhaps even the United States are willing to countenance more state participation in the oil industry at the expense of private enterprise.

Several years ago, when OPEC was selling only 6 percent of its oil (instead of the present 20 percent) on a government-to-government basis, a few so-called experts in the United States wanted the governments of the consuming countries, in particular the United States, to forbid buy-back agreements and to encourage more direct contact between OPEC and the governments of the oil-importing countries. It was thought that by offering inducements such as long-term contracts and quantity premiums, and perhaps by a liberal greasing of the right palms, it might be possible to get some of these oil-producing countries to chisel on the others since, according to such experts in these matters as Adelman, this is the natural tendency of "economic man" when participating in something so unnatural as a cartel. On the other hand, it never occurred to these experts that given the prevalence of enormous individual wealth in the oil-producing countries, as well as the greed of politicians in the consuming countries, the palms being crossed with silver—or gold—would most likely be those of their own incorruptible legislators and public servants.

Concomitantly, it was argued that when there was a glut of oil, direct contact between governments would make it easier for the producing countries to break ranks and begin cutting prices. There is a light glut appearing at the time I am writing these words, but instead of cutting prices, the producers are cutting production. From the point of view of their own economic self-interests, this is obviously the correct thing to do; and as these governments become more sophisticated, and the populations of the oil-producing countries become more politically aware, this kind of behavior will become a matter of routine.

To conclude this discussion, we can note that OPEC is not alone in trying to push the private oil companies further into the background. The Norwegian government has proposed that these companies no longer be allowed to acquire equity holdings in potential fields. Instead, they can become contractors for exploration at their own risk, which is not a particularly attractive arrangement for these companies, although it is ideal for

Statoil, the state oil company. Successful exploration will be rewarded with a share of the oil or gas produced, and the state oil company will reward these private companies with a bonus if production exceeds a certain level, although no indication has been given yet as to the proposed levels.

Ostensibly the Norwegian government desires to secure a firmer control over their output of oil and to be in a better position to determine prices and revenues. In particular, they want to stretch out the production of Norwegian oil to about one-hundred years. On the basis of OECD estimates of recoverable Norwegian reserves of 65 Gbbl, this means a production level of slightly under 1.8 Mbbl/d of oil. It is conceivable that this figure will be changed as more oil and gas are found, although it is impossible to say in which direction. As for the directors of the oil companies, they have reacted to this and similar proposals with a sense of resignation. Basically they feel that as production moves to more demanding water depths and regions, their services will prove invaluable. This may or may not be so since, in the long run, Scandinavia would be able to provide the technology to cope with any set of conditions, regardless of how unfavorable; and given the antipathy against multinationals in that part of the world, a future government might decide to take over all aspects of the oil industry.

We can now examine how the cost of oil to the oil companies is determined. As pointed out earlier, the petroleum going to the companies is composed of equity oil and buy-back oil. The price that an operating company pays for a barrel of equity oil is effectively equal to the cost of extraction plus the *royalty* per barrel plus the income tax per barrel. Since the producing countries are interested in only the net value that they obtain per barrel of oil exported, some question may arise as to why royalties and income taxes should be distinguished. One very good reason is that all or a portion of the taxes paid by operating companies to the governments of producing countries is deductible in their home countries.

The other component, buy-back oil, costs the operating company some percentage of the posted price. This posted price, as mentioned, is determined exogenously and is no more than a datum. The representative cost of a barrel of oil to a firm can be taken as a weighted average of the cost of equity and buy-back oil. To show what we mean, consider the following example. Take the posted price of petroleum to be $12/bbl, the total of production as $0.30/bbl, the royalty as 20 percent of the posted price, the income tax as 85 percent of the taxable base, and the price of buy-back oil as 90 percent of the posted price. Then we can make the following calculation:

	Posted price	$12/bbl
−	Production cost	0.30
−	Royalty (0.20 × 12.0)	2.40
=	Taxable base	9.30

	Income tax = 0.85 × taxable base	7.91
+	Royalty	2.40
+	Cost of production	0.30
=	Total cost of equity oil	\$10.61/bbl
	Cost of buy-back oil (0.90 × posted price)	\$10.80/bbl

The average cost of a barrel of oil when equity oil is 40 percent of the total amount produced and buy-back oil is 50 percent (and thus the producing country keeps 10 percent for selling through its own channels)

$$\frac{10.61 \times 0.40 + 10.80 \times 0.50}{0.40 + 0.50} = \$10.71/\text{bbl}$$

One final point needs to be looked at before we conclude this aspect of our discussion—the components of the price of oil products (such as gasoline, kerosene, heating and diesel oils, and residual fuels) sold by the oil companies in a major consumer country. One of the most complete breakdowns of this price is that of Chevalier (1974) for France in 1974. The total price of a representative unit of oil products is 820 francs/ton, and it is formed by the following components: Production cost (in producing country), 5 francs (0.5 percent); producing countries' profit 263 francs (32.1 percent); transportation (to France), 31 francs (3.8 percent); refining margin (in France), 25 francs (3.0 percent); distribution costs (in France), 35 francs (4.3 percent); taxes in France, 315 francs (38.4 percent); oil company's profit, 145 francs (17.4 percent).

It appears from these figures that in France, as in most of the major oil-importing countries, a large part of the price being paid by consumers for oil products consists of taxes. This is probably as it should be, since automobiles involve large social costs (for example, the cost of road upkeep and the salaries of traffic police) that do not devolve directly onto drivers; however, insofar as they express themselves on the subject, the producing countries object to tax systems that cause taxes to rise as the oil price rises. Nonetheless, I believe that in a situation where oil prices are rising at a fast pace, investments in energy-saving equipment becomes extremely important, and taxes on oil and oil products could help finance these and other investments.

Some Post-1973 Oil Politics

In the spring and summer of 1973, King Faisal of Saudi Arabia was hard at work trying to convince the United States to decrease its support for Israel.

Since contact through conventional diplomatic channels gave no results, he turned to the oil companies, informing their directors that the situation of these firms could hardly improve in regard to relations with their host governments if they did not use their influence in Washington to promote the message that he was trying to send the U.S. government. The oil companies duly interceded, but their entreaties fell on unhearing ears. The pompous bungling displayed by a portion of the oil companies' leadership had begun to bore the decisionmakers in Washington who, for the most part, had their hands full with the decline of the dollar and recovering from the Vietnam fiasco.

In August 1973, Sheikh Yamani announced that under no circumstance would Saudi Arabia increase production by more than 10 percent a year. Given the projections for the desired rate of economic growth over the coming year, it was immediately obvious that if these projections were to be realized, Saudi Arabia would have to alter its intentions. At almost the same time, Libya announced that it was unilaterally increasing the price of its crude to $6/bbl. This move caused President Nixon to initiate a futile attempt to persuade Colonel Khadaffi that he was making a mistake, reminding him, among other things, of the Iranian boycott of 1951-1954, when Iranian crude had virtually disappeared from the world market, and at the same time trying to convince his allies and well-wishers throughout the world that a little well-chosen rhetoric from the Oval Room of the White House could bring recalcitrant oil vendors to heel.

On October 6, war broke out among Egypt, Syria, and Israel. On October 8, at OPEC's regular meeting in Vienna, Sheikh Yamani demanded $6 for each barrel of oil that OPEC members turned over to the oil companies. The oil firms, in turn, bid $3.50/bbl. After a while, with OPEC's offer at $5/bbl and the oil companies' at $4/bbl, the meeting was adjourned.

The directors of the oil companies, still lingering in the dream that they counted for something where setting the oil price was concerned, claimed that they had to consult their governments before rendering a final decision. The OPEC directors, on the other hand, simply shuttled over to Kuwait where, without further ado, they fixed the oil price at $5/bbl. But now, with U.S. weapons pouring into Israel, the Israeli army across the Suez Canal, and the road to Cairo open, something more drastic seemed in order to the Arab members of OPEC. It took the form of a boycott of those countries which were most active in their support of Israel.

The United States and Holland would get no oil, while other countries on the pro-Israel list would have their deliveries cut by 10 percent to start and then 5 percent a month. (Those countries not on the list could still buy oil, but at the new high price, which was not much help.) It was this tightening of the oil screw that showed the consumers of the major industrial countries just what kind of world they were living in, rather than the possibility of a major confrontation between the United States and Russia in the Middle East, and the industrial countries reacted by making it clear that they would accept any arrangement that would keep their automobiles on the road. In November, OPEC met again in Vienna and informed the oil companies that since they were apparently short of constructive solutions to various problems of mutual concern, OPEC would take the matter of oil pricing into its own hands. Price formation via negotiation between producing countries and operating companies now belonged to history.

Shortly before the next OPEC meeting, OPEC economists advanced the opinion that an oil price of $17/bbl was called for by the existing state of supply and demand. This price can be tolerated now, but had it been imposed in the latter part of 1973 and maintained into 1974, would have disabled the economies of the noncommunist industrial world and probably brought on some kind of military intervention against OPEC. In the end, the price settled on was $11.65/bbl. As for the oil companies, they confirmed their new status by maintaining a deep silence. (By contrast, a few prominent members of the economics profession were able to enjoy a limited celebrity by concocting fairytales about how the new price was tolerable as long as a large part of OPEC's revenues could be "recycled" to the financial districts of New York and London.)

At present oil has started mixing with politics in a peculiar way once more. The Palestinian question has been actuated as a result of the Egyptian-Israeli peace treaty, and a number of oil-producing countries are making it clear that supplies of oil to the industrial countries will never be secure until the Palestinians have a homeland. If present trends continue, it seems likely that Israel will have to accept a Palestinian state on its border in return for some military guarantees from the United States (and possibly some others) of the type prevailing among NATO members; but whether this will mean that the industrial countries will obtain, at moderate prices, the oil *they* need to check rising unemployment and falling living standards remains to be seen. The basic prognosis at just this moment is for a declining physical availability of oil, and this situation cannot be changed as a result of either large or small alterations in the fabric of Middle Eastern politics. Last, but not least, there may be some problems in the offing for the major oil-consuming countries if the present tendency toward state-to-state deals intensifies and they find themselves having to negotiate with a few large countries whose actions are politically instead of commerically motivated. If the directors of Exxon or Shell were to take a dislike to some idiosyncracy of the president of the United States, they would find it not difficult but impossible to translate this dislike into a reduction of supplies to the U.S. market. But if the government of a producing country were in the same predicament, it could, if it chose, react immediately.

Oil and Economic Development

The first thing to clarify here is the theoretical purpose of an organization such as OPEC. Despite the opinion of large numbers of people in Europe and North America, its purpose is not to raise the price of oil as rapidly as possible, but to take what the members believe to be the correct steps to transform a large part of the oil assets of these countries into other forms of assets, particularly reproducible capital and an educational system capable of training people to cooperate with this capital in order to build societies with a high and sustainable material standard.

If this cannot be done within a reasonable time, the governments of these countries should be aware that when the oil bubble bursts—and burst it must someday—their countries will have the ill fortune to sink even further into underdevelopment than they are now. Some of these governments *are* aware of this fact, and they have started to think about just how much oil they are going to require in the distant future, particularly should their development plans, desires, and aspirations fail to materialize within a reasonable time scale, which could very easily happen. In addition, even if the development process runs smoothly, it seems certain that for many OPEC countries the first steps into full-scale industrialization should be via oil-based industries (such as petrochemicals), industries that are very energy-intensive (for example, the refining of aluminum), and a highly mechanized and energy-intensive agriculture. At least one OPEC country is sympathetic to this approach, since the government of that country has intimated that inexpensive oil will be available in return for petrochemical technology.

This brings us to the topic of just how much oil these countries will likely have in their possession when we look beyond the newspaper headlines and begin to think in terms of decades rather than months. Trying to predict the future is always a tricky business, but at this stage it is difficult to picture them having really impressive supplies. In fact, if we look fifty years into the future, it seems likely that of present-day OPEC countries, only a few in the Middle East will be important exporters of oil. This by itself would indicate an intensification in the reluctance of some of—but not all—these countries to part with ever-increasing amounts of what they will eventually perceive to be an invaluable national asset. It is in the sense of the above discussion that we can speak of oil as eventually having a greater value if left in the ground than if extracted and turned into cadillacs, color televisions, and first-class plane tickets to romantic places.

Now it has been claimed that only a few of the OPEC countries have the capacity to eventually construct non-oil-based economies. The theory here is that almost all the OPEC countries can build any type of society that they want, if they are capable of using their oil revenues to form that most valuable of all capital—human capital. This means not only schools, technical institutes, and universities, but also industrial projects where an industrial labor force, technicians, and managers can obtain the kind of industrial apprenticeship associated with modern, socially progressive societies. Everything considered, this means that OPEC countries (and other LDCs in a similar predicament) must initially concentrate on the development of intermediate industries whose integration into the local and international economy is facilitated by their link to existing industries.

Thus we come naturally to the refining of oil and the transportation of both crude oil and refined products. Between 1973 and 1977, OPEC's increase in refining capacity was less than its increase in domestic demand, and, excluding Venezuela, OPEC's exports of refined products are now only 5 percent of crude exports. Venezuela, on the other hand, can refine about 1.4 Mbbl/d, and if all capacity were utilized, Venezuela could refine 60 percent of the oil it produces. For a country in this situation, there is no technical impediment to eventually refining all its crude. Those who feel that refining offers no great promise for the OPEC countries argue that it involves huge investments having a small value added and only limited employment. Thus only a small part of the population might benefit from this activity, and the people doing so would tend to become a privileged class.

I believe that the worst industrial investments in the world, if they involve young people in purposeful industrial activities and thereby teach them correct work habits, are better than investment in British or U.S. treasury bills or apartment houses in the suburbs of Brussels or the center of Flaine. As for value added, a correct definition would include the profit of refiners, and as far as I know, few manufacturing activities are more satisfying from this point of view. In fact, the problem with going into refining usually has been uncertainty with respect to the supply of crude, and this is precisely the dilemma that OPEC countries avoid. Some figures can now be given showing world refining capacity in 1975 and OPEC countries' share of their domestic refining capacity in 1974. (See table 7-3).

Some of the exposition given above is found in a slightly different form in my book *The International Economy: A Modern Approach,* but one important point made there should be repeated here and extended slightly. If OPEC greatly increases its refining activities, it must accept a major financial risk in that, as things now stand, its refined products must be sold to the developed market economies. If the OPEC countries owned a large amount of the world refining capacity in the near future and there were a breakthrough by the industrial countries in the energy field (through the discovery of huge oil supplies somewhere in the world, the design of solar or nuclear equipment, or in the processing of coal or production of some type of synthetic oil), then OPEC would be put in the position of having to

Table 7-3
World Refinery Capacity, 1975, and OPEC Countries' Share of Refinery Capacity in Their Own Countries, 1974

Country	*Capacity (M tons)*	*Share (%)*	*Country*	*Share (%)*
Western Europe	1,042	29.0	Algeria	100
North America	874	24.5	Ecuador	13
Eastern Europe and China	610	17.0	Gabon	5
Latin America	381	10.6	Indonesia	100
Southeast Asia	144	4.0	Iraq	100
Middle East	139	3.7	Iran	100
Remaining	402	11.2	Kuwait	28
OPEC	192	5.4	Libya	87
			Nigeria	60
			Qatar	100
			Saudi Arabia	3
			Venezuela	2
			OPEC	40

Source: The Petroleum Economist (Various Issues).

watch a great deal of expensive equipment rust or else make drastic reductions in the price of both their crude and their refined output.

Moreover, if OPEC began moving into refining in a big way, the major oil companies would have an incentive to lead such a breakthrough. Some of the oil firms have already shown an interest in coal, and companies like Exxon and Shell would probably do better in the nuclear, solar, or synthetic-oil business than in publishing or processing nonfuel minerals. However, it has been suggested in more than one place that the world energy picture would be much more serene if the OPEC countries had a larger share of the refining and chemical industry. This would help them identify to a certain extent with producers in the industrial countries and therefore tend to make them more "responsible" in their pricing policies. This may or may not be so, but at the same time certain problems will be posed for the present owners of refining and petrochemical assets. Even the most successful European chemical countries are concerned about the threat posed to their sales by Middle Eastern products made from inexpensive ethane gas. Saudi Arabia in particular plans to become a force to be reckoned with on the world petrochemical market as a result of its copious supply of cheap oil-field gases. At present these gases are available to the Saudi Basic Industries Corporation (Sabic) for about $2/MBtu, which is about one-fourth what European chemical firms are paying and probably still less than what they will be paying by 1985. As for technology, this will be supplied by Mobil, Exxon, and Mitsubishi, who are the partners of Sabic in various petrochemical joint ventures.

The negative side of these projects shows up in the high capital cost of building chemical plants in the Middle East, where both infrastructure and local technical and administrative skills are operating at full capacity. Marketing will also be a problem, but not so much as some people think, since the sale of petrochemicals can to some extent, be tied to the sales of crude. Sabic plans to have a production capacity for methanol of 1,300,000 t in 1985, which would be 8.5 percent of the world supply. They also plan to produce large amounts of ethylene and ethanol (7.2 percent and 7.0 percent of the world supply, respectively). These are ambitious targets, and it will be interesting to see whether they are realized, or even partially realized, because *if* they are, it signifies an important, even a world-shaking, breakthrough on the development front: The ability of a less developed country (albeit a rich LDC) to mobilize, in less than a decade, the capital and skill necessary to challenge some of the industrial giants of Europe on their own turf. (Here it should be noted that this challenge intially concerns Europe, since the U.S. and Japanese companies which are included in the joint venture will probably be able to protect *their* home markets, at least in the short run.) There is also the possibility that later on projects of this type will be carried out in cooperation with some of the semi-industrialized countries of Asia, such as South Korea and Singapore, or even China.

Remarks similar to those above apply to the transportation stage, although on the surface it appears that the major oil companies are not particularly sensitive to the possibility of OPEC's making a large-scale intrusion into the tanker market. It has also been said that there is a palpable economic disadvantage in exporting refined products in comparatively small tankers rather than crude oil in supertankers. This may be so, although my own primitive calculations indicate that this disadvantage is more than offset by the profits on both transport and refining, especially the latter.

There is still, however, some excess capacity in tankers, and it will apparently take some time before this is cleared up. In 1976 the world tanker fleet came to 309 Mtons, with 56 percent registered in OECD countries and about 35 percent registered under "flags of convenience" (mostly Liberian and Panamanian). Japan owned 11 percent of the OECD ships (by weight); British shippers, 10 percent; Norwegian, 9 percent; and various Greek companies and personalities, 5 percent. The oil companies own 35 percent of the world tanker fleet, and thus an expansion of OPEC tanker capacity would not have to be exclusively at their expense.

According to a survey made in 1976, OPEC countries were in possession of 3 percent of world tanker tonnage, or about 10.2 Mtons. Of these Kuwait owns 2.1 Mtons; Libya, 1.5 Mtons; Saudi Arabia, 1.1 Mtons;

Algeria, 0.9 Mton; and Iran, 0.6 Mton. Given the present overcapacity in the tanker market, it seems unlikely that these countries will make any great effort to expand their tanker fleets in the near future. But suggestions have come from the government of Kuwait that it would be more appropriate if OPEC oil were carried on its own tankers; and we will probably be hearing more of this in the future.

One topic remains. Although the latest figures are not available, OPEC's total foreign assets were approximately $300 billion at the beginning of 1980. These must be considerably larger now because of the very steep increases in the price of oil during 1979-1980, and because the ability of several of the major oil exporters to absorb their oil revenues in the form of goods and services is still limited. Table 7-4 shows the OPEC foreign-asset holdings as of October 1978.

Many OPEC countries are thinking in terms of building up more sophisticated banking systems that can invest more money locally. This is particularly true in the Middle East where, during 1979, five Islamic banks were opened in Kuwait, Iran, Bahrain, Jordan, and Egypt. One of the new ideas associated with some Middle Eastern banks is that customers will have an equity position with their bank. There is also a broad cooperation with banks in the OECD countries: London and Paris have about fifty financial institutions in which there are Arab interests. In addition, banks such as the Saudi International Bank, European Arab Bank, and Union Banque Arabe et Française are involved in consortia with a number of large U.S., European, and Japanese banks.

The Rotterdam Spot Market

Rotterdam, Holland, is a huge refining center and the largest petroleum-handling port in the world. Conceptually, however, this has nothing to do

Table 7-4
OPEC's Total Foreign Assets in 1978
(billions of U.S. dollars)

Investments in the United States	45
Investment in industrial countries other than the United States (mainly Germany, Switzerland, France, Japan, and Canada)	25
Direct loans to industrial countries (including the United States)	13
Deposits on the Eurocurrency market	70
Bilateral aid, loans to the LDCs, and investments in LDCs	22
Loans to international financial institutions	10

Source: Arab Monetary Fund Annual Reports.

with the Rotterdam spot market, since the spot market is a way (rather than a place) of doing business. For the most part, oil is sold in bilateral deals between buyer and seller for future (forward) delivery. The exact price and quantity can be fixed in the contract, but it could *also* be stipulated in the contract that the price paid by the seller at the time of delivery be other than that prevailing at the time when the contract was drawn up. For instance, if the posted price of oil increases from the time at which the contract is entered into until the date of delivery, then the buyer might have to pay an appreciably higher price to get the quantity of oil specified on the contract. It may happen, however, that the purchaser of oil underestimates or overestimates his requirements when placing his contracts. If he underestimates, then he purchases the shortfall on the *spot* market (which is also referred to as the "free" market, where price is set by short-run demand and supply). Obviously, if the buyer is in Marseille, his "spot oil" need not come from Rotterdam. Instead it might arrive via a tanker that was in the Mediterranean and was redirected toward France when he gave his order (although, in such a case, there is a strong possibility that his order was handled by a Rotterdam oil broker).

Similarly, when buyers have too much oil, this oil can be sold on the spot market. There are other possibilities here. In Sweden, several years ago, when it appeared that a small oil glut was in the offing, buyers placed fewer orders than usual with their customary suppliers (the large oil companies and the state companies of the producing countries). The glut did develop, and since a great deal of oil was sold on the spot market by various countries and organizations, the spot price of oil fell rapidly and many Swedish buyers made some lovely profits. On the other hand, when the price of oil was rapidly escalating, in 1979, and buyers over the entire world were filling their storage tanks to the brim against the possiblity of future shortages, the Swedish buyers who had not covered their requirements through long-term contracts had to pay very high prices for spot oil. Note the dilemma faced by both buyers and sellers when a commodity can be bought and sold on a forward market, where prices and quantities are specified, and/or on a spot market, where price formation depends on short-run supply and demand conditions: For example, should a buyer buy forward and thus be assured of his requirements, or should he take a chance and wait to buy on the spot market, hoping that the price will be low? In my courses in mathematical economics I have often treated this kind of problem, and as far as I am concerned academic economics can offer no solution for the general case, despite claims to the contrary.

Of late not only are the oil-producing countries selling more oil directly to the importing countries, but also they have raised their deliveries to the spot market. Present estimates are that these countries now sell about 15 percent of their output on the spot market, as compared to 10 percent during 1978 and slightly less earlier. The increased deliveries began in the middle of 1979 when the spot price of oil increased very rapidly relative to the posted price. Under the circumstances it made economic sense for anyone who was in position to do so to divert supplies into the spot market. At the same time, though, it should be apparent that sellers generally would be unwilling to raise the amount sold on the spot market indefinitely because, in doing so, they would have to accept the greater amount of uncertainty associated with the spot as opposed to the forward market. As bad luck would have it, there is not much room for textbook behavior in situations characterized by the kind of uncertainty existing on the world oil market. It could be argued though that the problem is with the textbooks, and not with the markets.

Oil and the Third World

The agonies which the oil price rises caused the industrial countries have been compounded in the Third World. This is true of all categories of countries. Even the brilliantly successful South Korea has seen its growth dramatically slowed as energy costs increase. That country's oil bill may be as large as $6 billion in 1980-1981, although the total value of goods and services produced will probably not exceed $62 billion. The trade deficit has also reached a record level, because no matter how efficient the South Korean economy may be, it has no possibility at all of paying for its oil imports out of current earnings. Instead the South Koreans trade deficit is financed by loans from the United States, Europe, and international organizations such as the World Bank.

The current-account deficit of oil-importing LDCs, the petroleum import bill for these countries, and the amount of aid received in several years are shown in the table 7-5. What this table does not indicate is that each percentage-point reduction in the growth of the major nonsocialist industrial countries reduces the exports of these countries by $2 billion, which implies additional, and perhaps insurmountable, difficulties in trying to service the huge amount of debt accumulated by these countries.

The question might be put as to whether there is any way out of this predicament for these countries. The simple truth is that there are two, both of which are certain winners. The first is discipline—discipline in education (that is, educating the right people for the right occupations), in population growth, and in administration. The second has to do with culture. The less developed country that has done the most with the fewest resources is Singapore, which tailor-made its culture to suit its goals. For example, although they are a multiracial society (Chinese, Malayan, Indian, and Tamil), they have adopted English as the official language of the country, as well as a low threshold of tolerance for some of the more unsavory

Table 7-5
Current-Account Deficits and Oil Import Bill for the Oil-Importing LDCs, 1973-1990

	1973	*1975*	*1977*	*1979*	*1980*[a]	*1985*[a]	*1990*[a]
Current-account deficit (at current prices and in billions of U.S. dollars)	−6.2	−39.0	−24.3	−43.2	−63.3	−72.9	−85.1
Current account deficit as percentage of GNP	1.0	5.1	2.6	3.3	4.2	2.7	1.8
Petroleum import cost (in billions of U.S. dollars)							
Low-income countries[b]	0.6	1.8	2.0	2.7	3.3	6.0	11.1
Middle-income countries[c]	6.1	20.3	26.1	40.4	54.5	101.2	186.9
Total	6.7	22.1	28.1	43.1	57.8	107.2	198.0
Foreign aid[d]	13.0	—	—	29.0	32.0	—	—

Source: World Bank documents; GATT documents; IMF documents.
[a]Estimated.
[b]Countries with a per-capita annual income of $300 or below.
[c]Countries with a per-capita annual income greater than $300.
[d]In billions of U.S. dollars.

"freedoms" of the capitalist world. Still, its economy is unabashedly hard-line capitalist and always has been—even before the prime minister of Singapore was drummed out of the socialist internationale.

Unfortunately, for most LDCs, both remedies are out of the question, and neither the politicians who run these countries nor the ladies and gentlemen of the aid and international organizations would dare suggest them in private, much less in public. Instead they insist on the "massive transfer" of financial resources, to use a particularly badly chosen phrase from the report of the Brandt Commission. I believe, though, that these transfers are already too large. They should be replaced by trade arrangements which involve removing all restrictions against Third World manufactures and semifabricates.

Appendix 7A: A Mathematical Comment on the Income of Oil Producers and a Comment on Quotas

The problem broached in this appendix was suggested by the work of Calvo and Findlay (1978). They are concerned with the relationship between oil export policy and the return on investing in the capital stock of the oil importer. I am concerned with maximizing the income of oil exporters. Put more humbly, I am concerned with formulating an algebraic exercise which illumines some of the issues that should interest economic theorists considering this category of problem. Note my clear reference to economic theorists, and not applied economists, wildcatters, oil millionaries, or the International Energy Agency, since under no circumstances do I labor under the misapprehension that exercises of this type have any profound practical significance.

Having said that, I can proceed with a little algebra. My starting point is the basic Harrod Domar growth equation:

$$sY = \frac{1}{v}\frac{dY}{dt} \tag{7A.1}$$

Here s is the (marginal and average) savings coefficient, and v is the marginal output-capital ratio. If there is an intermediate input, that is, oil, then Y is gross aggregate output, and the income of residents of the country is not Y but $Y - px$, where p is the price of oil in terms of the composite good represented by Y and x is the amount of oil necessary to produce Y. Now, as usual, x can be determined via an input-output relationship with Y: $x = \Theta Y$. Thus we have

$$s(Y - px) = s(Y - p\Theta Y) = \frac{1}{v}\frac{dY}{dt} \tag{7A.2}$$

or

$$\frac{dY}{dt} = svY(1 - p\Theta) \tag{7A.3}$$

We want to maximize the discounted revenue of the oil exporter, or

$$\max R = \int_{o}^{\infty}(p\Theta Y)e^{-rt}dt \tag{7A.4}$$

where r is a discount factor. The continuous-maximum principle suggests formulation of the following Hamiltonian:

$$H = p\Theta Ye^{-rt} + \lambda[svY(1 - p\Theta)] \qquad 1 \geq p\Theta \geq 0 \qquad (7A.5)$$

Here p is the control variable, and Y is the state variable. Differentiating partially with respect to the control variable, we get

$$\frac{\delta H}{\delta p} = \Theta Ye^{-rt} - \lambda svY\Theta = 0$$

or

$$\lambda = \frac{e^{-rt}}{sv} \quad \text{and} \quad \frac{d\lambda}{dt} = -\frac{re^{-rt}}{sv} \qquad (7A.6)$$

With respect to the state variable,

$$-\frac{\delta\lambda}{\delta t} = \frac{\delta H}{\delta Y} = p\Theta e^{-rt} + \lambda sv(1 - p\Theta) \qquad (7A.7)$$

Combining equations 7A.6 and 7A.7, we get

$$r = sv \qquad (7A.8)$$

This would appear to be a dead end, since the variables have entered the formulation in such a way that we did not get the variable we were interested in, p, to appear in our final expression. We could reformulate the problem to clarify this matter, since the key issue here turns on the convergence of the integral in equation 7A.4, in conjunction with the specification of an infinite time horizon. But without considering these matters, let us reformulate equation 7A.4 slightly:

$$R = p\Theta Y_0 \int_0^\infty e^{[sv(1-p\Theta)]t - rt} dt$$

$$= p\Theta Y_0 \int_0^\infty e^{[sv(1-p\Theta) - r]t} dt \qquad (7A.9)$$

This expression is for cumulative exporter revenue over an infinite time horizon. Now let us use our result $r = sv$. Then we get

$$R = p\Theta Y_0 \int_0^\infty e^{-svp\Theta t} dt = \frac{Y_0}{sv} = \frac{Y_0}{r} \qquad (7A.10)$$

This simply indicates that if the discount rate equals the natural rate of growth (in the Harrod-Domar sense), then the asset represented by the capital stock of the importing country yields a discounted revenue whose gain due to growth is exactly offset by the disadvantage of having to wait for this gain. Now suppose we assume that $p\Theta = 1$, which is a situation where all output is taken from the importing country is return for their oil. In this case we have

$$R = Y_0 \int_0^{\infty} e^{-rt} dt = \frac{Y_0}{r} \tag{7A.11}$$

The common sense of this is obvious. In every period the oil exporters take all the importers' revenue. If we assume no depreciation, this takes place for an infinity of periods, and so discounted revenue is Y_0/r. This is as in equation 7A.10. Now, to cut a long story short, if in equation 7A.8 we have $sv(1-p\Theta)<r$, then the integral converges and we get by integrating

$$R = \frac{p\Theta Y_0}{r - sv(1-p\Theta)} \tag{7A.12}$$

If we differentiate this with respect to p and set it equal to zero, we get the same result as in the optimal control problem. Thus we need a numerical solution with given values of r, s, v, and Θ, different values of p can be plotted against R and the maximum can be noted. What about when $r>sv(1-p\Theta)$? In this case the integral is not convergent, but if we choose a situation where the time horizon is finite, a solution might still be possible.

In this chapter an import quota put on oil by President Eisenhower was reviewed briefly. Here we can take up some of the technical details of that quota with the aid of figure 7A-1. In the situation shown in figure 7A-1*a*, with domestic demand d U.S. suppliers provide q_d, and importers $q_t - q_d$. With a quota on imports of q^*, the amount supplied by importers is shown in figure 7A-1*b* as AB ($= CD = EF$) $= q^*$. Imposing this quota raises the price of oil from p' to p'', because the quota changes the effective supply curve from $s_{us} - s_T$ to $s_{us} - A - B - D - s''_{us}$, and with the same demand curve that we had in figure 7A-1*a* the supply-demand curve intersection is at D. As postulated, foreigners supply $CD = q^*$, and total production is now q''_t, with domestic production $q''_d = q''_t - q^*$. Obviously producers in both countries have won this arrangement, while consumers have lost (unless they work for oil companies). A problem has been created here, however, as to who are the lucky foreigners that get to sell oil at p'' which they were previously willing to sell at p'. In the case of the U.S. import quota, foreign producers were allowed to make sales on the basis of their share of the U.S. import market the year before the quota went into effect.

Making a decision of this kind could result in corruption. Since foreign producers now have excess profits of $(p'' - p')q^*$ on sales of q^* (if we assume that earlier they were just covering costs, to include normal profits), some of these profits are available to bribe the officials passing out import licenses. This is common practice in some countries and explains why some economists prefer tariffs to quotas.

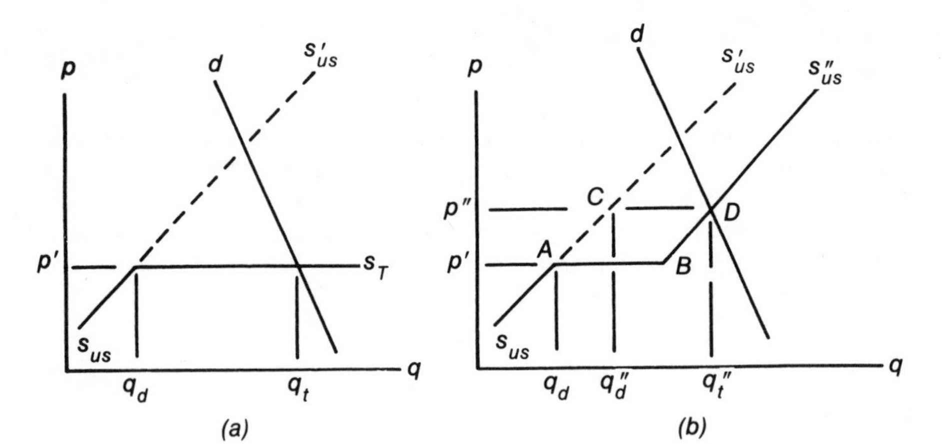

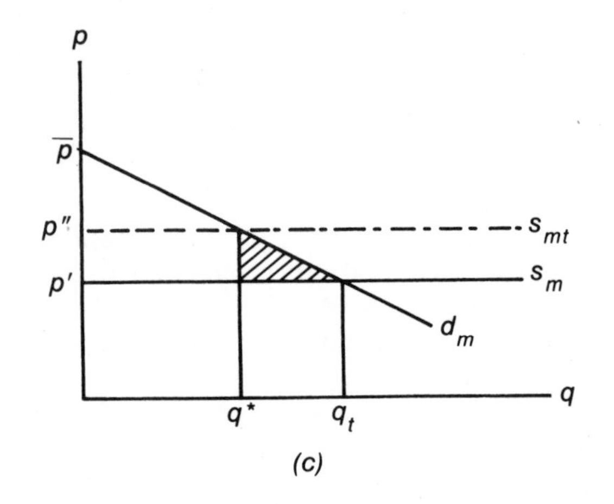

Figure 7A-1. Supply and Demand Curves for a Country Importing Oil, before and after the Imposition of a Quota

Our discussion can also be carried out by using a slightly different diagrammatical apparatus. Following the analysis of Caves and Jones (1977), the supply and demand curves s_m and d_m, for imports are shown in figure 7A-1*c,* where domestic supply is omitted. Note that the price $\overline{p}$ corresponds to the price resulting from the intersection of d and s_{us} in figure 7A-1*a*. Imposing the quota q^* (smaller than the free-trade foreign supply q_t) drives up the domestic price to p''. This quota causes a price distortion of $(p'' - p')/p'$ and thus is equivalent to a restrictive tariff of this same amount. Thus s_{mt} becomes the effective-supply curve for imports. Here we see clearly that had the domestic government used a tariff instead of a quota, it could have realized an income of $(p'' - p')q^*$. In the case discussed in this chapter the U.S. Congress refused to authorize a tariff (or a subsidy to U.S. oil producers), and so foreign sellers of oil (who in some cases might have been Americans) captured most of or all the $(p'' - p')q^*$. The welfare cost of this type of restriction is shown by the crosshatched area.

Summary, Overview, and Conclusions

This book consists of a long excursion through the economics of the world oil market, to include related phenomena of both a macroeconomic and a microeconomic nature. No space need be wasted here to bring home the seriousness of this topic. In the preface I make what I think is the most important point in the book. The three absolutely crucial topics in economics today are the economics of population, the economics of nonfuel minerals, and the economics of energy—*in that order*. If the world population can be stabilized at a reasonable level, it goes without saying that the energy problem can be solved, since renewable energy sources and innovations such as fusion seem to offer great promise in the long, but probably not the short run.

But if things continue to develop as they are now, the world population is *not* going to be stabilized at anything near a reasonable level. The demographic cards for the 1990s have already been dealt, with the result that hundreds of millions of people in the LDCs, or perhaps even billions, are going to leave the game empty-handed, regardless of what the overpaid and overfed café philosophers venting their impressive verbal skills at one or another United Nations talkathon have to say about the matter. What about the early twentieth century? People who are supposed to know about these things are crowing that world population about the year 2000 will be closer to 6 billion souls than the 7 or 8 billion they had once feared. What these experts are not saying, though, is that this 6 billion is a stable population. In fact, this 6 billion figure will be even more heavily weighted in favor of the most fertile age groups than today's rapidly growing population, and instead of stabilizing may be on the brink of an escalation that could bankrupt the carrying capacity of this small planet's ecosystem.

If we assume that such things as solar and wind power, fusion, biomass, and so on are economically feasible, the last part of this century and the first part of the next should be featuring a systematic conversion to these and other energy technologies. Note carefully, however, that this conversion may just be beginning and could require decades before it is completed. On the other hand the already huge, but rapidly increasing, world population will exhaust the remaining traditional energy supplies at a record pace and, by driving up the price of these supplies, will reduce real incomes in such a way as to have a catastrophic effect on the ability of most industrial countries to finance a new energy technology. As I point out in both this book and elsewhere, this is precisely the dilemma facing many Third World countries today, and some of these countries are reacting by visiting irreversible ecological damage on their immediate environments in an attempt to obtain energy materials that are essential for their basic needs. But when, or if, the industrial countries are put in this position, the damage most likely will be visited elsewhere and probably on a scale unimaginable since World War II.

In chapter 2 I refine and extend some arguments that I introduced in my books *Scarcity, Energy, and Economic Progress,* and *The International Economy: A Modern Approach.* At present about one-eighth of U.S. exports and 25 percent of Swedish exports are required to pay for these countries' imports of oil, where the *Swedish* figure should be put in perspective with the 7 percent that was required at the beginning of the 1970s. This trend will most likely continue, so oil will become an even more important part of the foreign sector of most industrial countries than it is today. It will dominate imports, and a growing percentage of exports will have to be given up for oil *and* for the raw materials and various intermediate goods needed to produce the exports which will pay for oil. In order to raise private consumption, extensive foreign borrowing will be required, although many countries already have a nearly intolerable foreign debt. What must be understood here is that increases in the price of oil function as a consumption tax on the oil-importing countries, and the only aggregate adjustment that is possible is to adjust to a lower standard of living, sooner or later. Also, for reasons alluded to below, the export industries of the oil-importing countries cannot possibly be expanded at the rate at which the oil price can be raised.

However, as is pointed out in chapter 2, nothing would be solved for the oil-importing countries if this were possible. While an increase in exports of goods and services to the oil-producing countries can be regarded as a triumph for an individual firm, it can easily entail a sacrifice for the country in which the firm is located. Take a situation where a firm in a country with a shortage of hospital facilities contracts to build and equip a hospital in some oil-exporting country. This may mean profits for the owners of the firm, bonuses for its workers, and promotions for its engineers; but this is hardly a consolation for those sick people in the oil-importing country who need extensive medical attention. If, for example, they have gout and look forward to being cured so that they can resume their champagne diets and get back to making the rounds of their favorite discos, they are unlikely to be exhilarated by the shouts and huzzahs of technicians and hardhats on their way to dance, drink, and carouse in the Gulf. The basic dilemma here is that there is a burgeoning discrepancy between the amount of real resources that has to be surrendered for a barrel of oil and the *productive power* of a barrel of oil as an industrial input (or, if the reader prefers, its utility in a consumption activity such as motoring). However, as argued in chapters 2 and 3 of this book, the rather special economics of an exhaustible resource may necessitate discrepancies of this kind, and it might not be appropriate to attempt to alleviate them with antitrust legislation or government regulation even if this were possible—which, in the international setting of the world oil market, is usually not the case. Instead, about the only thing that governments can do in the short run is to design insurance and transfer schemes that permit an equitable sharing of the ordeals that will be imposed by large increases in the price of energy, rather than permitting them to be vaulted over onto a few groups in the community.

Although a portion of the previous exposition was, for pedagogical reasons, delivered in a frivolous tone, there is nothing frivolous about the subject. One of the main purposes of this book is to impress upon the reader the *real cost* of the oil price rises, and this requires looking behind monetary phenomena at the allocation of real resources. At this point a simple numerical example might be useful. Let us say that 1,000 workers go to the Gulf to build a 1,000-bed hospital, and they receive for this work 1,000 bbl of oil which, in their home country, is used as an input to produce 1,000 chipped beef sandwiches. Notice that in this example there is a *primary factor* of production, labor, which produces one *final product*, hospital facilities, which are exported to the Gulf in return for oil. This oil is used as an *intermediate good* in the production of another final good, chipped beef sandwiches. To keep things simple, let us assume that there are no other costs in the production of hospitals or sandwiches. Prices in real terms are also of some interest. One barrel of oil costs one hospital bed (or place), and one chipped beef sandwich costs one barrel of oil. One hospital bed, in turn, costs the labor of one worker over a certain period. It also seems reasonable to assume that the option is always open for the oil-importing country to produce some hospital facilities at home; but with a fixed amount of productive resources (in the form of labor) this would mean getting fewer barrels of oil and thus having to reduce the output of chipped beef sandwiches in the home country, where historically we shall assume that 1,000 sandwiches were regarded as almost a biological *and* psychological necessity.

What about ten years ago? Ten years ago the same firms sent 100 workers to the Gulf where they constructed a 100-bed split-level medical center in return for 1,000 bbl of oil (which, as today, was duly transformed into chipped beef sandwiches). The other 900 workers stayed home and assisted in the less glamorous work of caring for and rehabilitating the numerous gout cases that seem to appear in the home country after each grape harvest. But observe the "productive power" of a barrel of oil ten years ago from the perspective of the home country! It permitted the production of not only a chipped beef sandwich, but also some medical services. Note particularly that the employment of production factors (labor services) in the home country was the same as today, but production (and consumption) was greater because it consisted of the same amount of sandwiches as today in addition to medical services. For this reason we can unequivocably state that real income at that earlier date was higher. We also see that the price of a barrel of oil at that time was only one-tenth of a unit of labor, and this was also the price of a sandwich.

And what will be the situation ten years in the future when, according to the latest forecasts, one barrel of oil will cost two hospital beds, or two labor-hours? Under these circumstances 1,000 workers can (indirectly) produce only 500 chipped beef delicacies. They are still employed, for which they are probably grateful, but they are working more and more for less and less. Here it is possible that we have the initial stages of what Nicholas von Hoffman has termed *full-employment poverty*. What about the assumption that traditionally 1,000 sandwiches were considered a biological and psychological minimum for the home country? There are several ways of handling this problem, at least the psychological part. The most refined is to get people to think they are as well off when they are consuming 500 sandwiches as when they were consuming 1,000. One way to do this is to provide them with another commodity—*money*. Thus the government might not oppose a steady increase in money wages since, although money buys less, people can at least get some pleasure from looking at their cash, fondling it, and counting and recounting it.

Are there any positive steps that can be taken to get out of this situation? One suggestion that will be offered involves training or breeding more people who can produce hospital facilities. This would lead to more exports to the Gulf, which in turn means more oil, which in turn means more chipped beef sandwiches. But training or breeding people is a serious business, and it takes both time and other scarce resources. On the other hand, many of us have the feeling, which is undoubtedly correct, that on some (but not all) occasions the oil price was summarily raised in an ambience of yawns and clinking coffee cups, more or less arbitrarily, and without the slightest thought as to its ultimate consequences for buyer or seller. As far as I am concerned, for the oil-importing countries as a whole (though not, perhaps, for a few exceptions) it is a singularly dangerous waste of time to try to counter oil price increases with greater exports. The only sensible economic solution is massive conservation and the installation of a new energy technology as soon as possible. On several occasions, incidentally, I have heard this solution proposed by various OPEC officials.

The reader can elaborate on the above example as he or she sees fit, because there are definitely some aspects of this issue left to explore in either verbal or numerical terms but one caveat is in order. At the heart of the previous analysis is the little-understood concept that the oil-importing countries are negatively affected by oil price increases when they must accommodate larger expenditures by the oil-exporting countries. This idea is spelled out in detail in the second chapter of *Scarcity, Energy, and Economic Progress*. By way of contrast, the oil-importing countries could show a net *gain* in their dealings with the oil-exporting countries if the latter do not spend any of their revenues, but "recycle" (that is, lend) them to the oil importers, who then proceed to invest them in such a way that the marginal return (or marginal yield) of this recycled money is at least as great as the real rate of interest that is paid for it. Unfortunately, this is easier said than done. As far as I can tell, although a considerable proportion of all revenues have been recycled to the industrial countries, they have not been invested. On the contrary, they have been and are being consumed, thereby contributing to the widening gap between debt and productive capacity which is guaranteed to provide a large number of people in the future with a burden that they will not be happy to assume—particularly those who were not in a position to enjoy the earlier feast.

Before continuing to more serious matters, I want to make it clear that there is nothing disrespectable, at least in economic theory, about borrowing from the future, either one's own or that of one's children. Until recently it was sound to assume that the future would be more affluent than the present, and a great many economists still take that position. Certainly, in the interests of equity, it might have been a good thing if some of the tremendous wealth created in the United States in the 1950s and 1960s had been available for U.S. citizens in the 1930s. However, in my opinion neither the growing unemployment nor the high rate of inflation that characterizes the industrial world today can be reversed in the near future. Both originate in the oil price rises and the subsequent transfer of real wealth to the oil-producing countries, and this transfer could easily accelerate. Even more important, in many countries the "cake" has virtually stopped growing, and thus gains made by one group will have to be paid for by another. This can hardly be conducive to overall social harmony, which constitutes an important component of the standard of living.

The most important topic dealt with in chapter 2 is the effect of the energy price rises on investment. These price rises not only have made some capital equipment and durable consumer goods obsolute in the sense that their operation is uneconomical, but also, by raising the operating costs (or, equivalently, lowering the yield) of various prospects, have caused their acquisition to be either delayed or canceled. In the United States the growth in output per worker-hour that characterized the pre-1973-1974 period has drastically slowed as a result of a deceleration in the process of substituting capital for labor. From 1948 to 1973 a high rate of investment caused a growth in the capital-labor ratio of about 3 percent a year, but from 1973 that rate has fallen to 1.75 percent because of low rates of investment. As a result, it has been said that even under the best of conditions it may take at least a decade for U.S. productivity growth to resume its earlier pace. It is also clear that the oil price rises have been one of the main contributory factors to inflation. They have increased the cost of both consumer and industrial activities and, by affecting investment and productivity in the manner implied above, raised future inflation rates above what they otherwise would have been. In addition, the oil price rises have been the direct cause of the depreciation of the currencies of various countries, which has contributed to an elevation of their general price level. According to the OECD, each 10 percent increase in oil prices leads to an average 0.5 percent rise in consumer prices in member countries (and also an average gross national product reduction of 0.3 percent). A few countries can manage this type of strain, but most cannot. Certainly Italy, Britain, France, and the United States—which already have high levels of unemployment, cannot resort for very long periods to the restrictionist policies usually employed against inflation (although certain politicians in these countries are apparently determined to use them anyway).

The significance of investments also needs to be emphasized. In both simple algebraic terms and historical experience, investment leads to economic growth and higher living standards for all groups in a given community. By initially shifting resources out of consumption, the Japanese have invested a much larger fraction of their output than the United States and the other industrial countries, and the Japanese have been rewarded for their foresight by an increase in growth *and* productivity that has been still larger proportionally. In fact, in slightly more than twenty years the level of Japanese technology has come abreast of that of the United States, and today it is moving ahead. In *Scarcity, Energy, and Economic Progress* I examined the achievements of the Japanese steel industry up to 1977. Its latest exploits are even more spectacular, if that is possible. By investing in the most modern automatic handling equipment and computer-controlled blast furnaces, the Japanese can outcompete, on any basis, any steel producers in the world. The productivity decline featured in many sectors of the U.S. steel and automobile industries has been avoided in Japan because productivity bottlenecks have been broken by the installation of new, automated equipment that virtually eliminates human error. The basis of this accomplishment is the cooperation forged between the Japanese government and Japanese industry, which has worked to mitigate those factors, such as energy price rises, that have tended to suppress investment in various industries in the United States and other countries. Tax incentives have been provided which stimulate productive investment, and special attention has been paid to making sure that the recovery of capital expenditures is not negatively influenced by inflation. Equally important are the more or less tacit provisions made to eliminate uncertainty about the future availability of finance.

In order to understand the significance of the last sentence in the above paragraph, the reader should try to comprehend that the energy price increases have resulted in the entry of uncertainty into economic matters in a manner and depth virtually unknown in previous economic history. One of the most significant results of modern economic theory (which as yet is almost completely ignored by the writers of elementary and intermediate textbooks) is that in the presence of uncertainty resources can be "correctly" allocated by the price mechanism *only* if we have a complete system of futures and insurance markets. With only a few execeptions, such markets do not exist at present, nor is it conceivable that they will become available in the near or distant future. Certainly this is true of energy markets, with their dependence on international cartels and the state of geological knowledge.

In order to make it easier to follow some of the issues being discussed here, mention can be made of the plight of Dutch consumers of gas who are now buying foreign gas at a price higher than that at which Dutch gas is being exported to such affluent countries as West Germany and Switzerland, among others. The reason for this strange situation is that in the 1960s the directors of the company exploiting the huge Groningen natural-gas deposit, *acting on information available at that time*, were concerned about not being able to sell their gas in the long run in competition with such things as oil and nuclear power. Long-term contracts were therefore entered into with many buyers, some of which called for deliveries during a twenty-five-year period. Thus, after the oil price rises of 1973–1974, a large amount of Dutch gas, because of contractual obligations, was still being sold at prices well under the international energy price; in addition, a precious national energy resource was being used up at an unacceptable rate. Texbooks of the future will undoubtedly have a great deal to say about this type of situation, but one observation can be advanced now. No safety mechanism can possibly be devised whereby the private economy can deal with uncertainty in the above form unless the art of reading crystal balls or tea leaves takes a quantum jump forward. Recognizing this fact, the Dutch government has threatened to cut off supplies to the purchasers of Groningen gas unless they declared themselves willing to renegotiate existing contracts.

The principal topic taken up in chapter 3 is the turning down of the world oil production—specifically, its turning down while more than half of the world's stock of recoverable oil is still in the ground. The key to understanding this phenomenon is comprehension of the so-called economic rate of recovery. For economic reasons (which, as I will show below, are associated with the amount of ultimately recoverable reserves in a given field), no more than a fraction of a given oil field's reserves should be removed in any given year. This fraction is generally put at about one-tenth, although it can vary somewhat from this figure, depending on the geological characteristics of the particular field. If, for example, we have a field containing 150 units of oil and we desire to lift 10 units a year, we can do so without violating our criterion during the first six years. In these years the reserve-production (*R/P)* ratio falls from 15 to 10 (that is, 150/10, 140/10, 130/10, . . . 100/10), as reserves fall by 10 units a year. But after the sixth year, if we produce 10 units per year, the *R/P* ratio becomes less than 10: that is, we would be lifting more than one-tenth of the field's reserves in a year. Thus, from the sixth year on, the *R/P* ratio "takes over" and determines production.

In the sixth year, reserves were 100 and production was 10. In the seventh year, reserves are 100 − 10 (=90), and with a desired *R/P* ratio of 10, production is limited to 90/10 = 9. The following year, with reserves equal to 81 (= 90 − 9), production would be 81/10 = 8.1, and so on. The reader should also observe that this example was carried out for a given field. In a typical oil-producing country, there would be fields which would be under development and not producing oil, even though they contribute to the total proved reserves. The same thing is true for the world as a whole; as a result, an appropriate worldwide *R/P* ratio is considered to be 15. In the United States, including Alaska, the *R/P* ratio is now under 10, and in the lower forty-eight states may be under 9 and close to what geologists consider to be the absolute lower limit.

The thing to notice in this example is that when production "turned down," there were still 90 units of reserves in the ground out of an original 150, which is well over half of the original reserves. At present the consensus is that there is about 2 trillion barrels (Tbbl) of exploitable oil in conventional locations, of which 400 billion barrels (Gbbl) has been produced, and about 600 billion barrels (Gbbl) falls in the category of *proved reserves* (or oil that can be economically produced with the existing technology). The other 1 trillion has yet to be discovered, but so-called informed opinion is almost certain that it will eventually become available.

In contrast to the last numerical example, the world demand for oil is not static, but is increasing at a rate of about 5 percent a year. As a result, world oil production could peak in the late 1980s or early 1990s with as much as 1 trillion tons of recoverable oil still in the ground. It has been said, however, that even if world demand increases by just 2.5 percent a year, production will still turn down early in the next century. Here it should be noted that U.S. production has already peaked, and, for the first time in history, the drilling boom that followed the oil price rises of 1973-1974 did not locate appreciable amounts of oil in that country. In connection with this discussion, however, one point should be emphasized. The critical *R/P* ratio (for example, 10 or 15) is basically an economic rather than a technical concept. True, it is technical in that too rapid depletion of an oil field physically damages the field, and this reduces the amount of oil which ultimately can be taken from a given field. But the most important aspect of this situation is that this reduction in reserves reduces the expected (discounted) cash flow from a field. By extension, in an emergency (such as a war) the *R/P* ratio can be pressed down to well under the critical value. It has been said, for example, that in a few years Russia will have to import oil, rather than continue as an exporter. This matter is discussed at greater length later in this chapter, but here it can be said that with oil prices at their present level (and climbing), an energy-rich country such as Russia, which expects important additions to its energy potential in the not-too-distant future from such sources as coal and the Siberian oil and gas deposits, might find it not only expedient but economically sound to allow the *R/P* ratio to fall well under the critical value for a few years in order to finance these new energy sources.

By way of contrast, unseemly high *R/P* ratios, in some instances, can also be justified. In 1979, at a meeting of the Institut für Wirtschafts und Gesellschaftspolitik, Dr. Mahjoob Hassanain of OPEC's economic secretariat stated that Saudi Arabia was not interested in having its *R/P* ratio drop below 20 to 25. Although Dr. Hassanain did not go too deeply into the matter, this remark indicated that the government of Saudi Arabia understands that during the rest of this century it will hardly have begun to replace this invaluable wasting asset (oil) with the right kind of industrial and educational assets, and thus, in their opinion, an *R/P* ratio lower than 20 could jeopardize their long-run economic development. [They are not, for example, going to produce anywhere in the neighborhood of the 20 million barrels per day (Mbbl/d) that the American Arabian Oil Company (Aramco) once gave as a production goal.] This is a complicated matter, and it has been elaborated earlier in this book, but it comes down to the following. If OPEC oil has alternative uses in the distant future (as, for example, an input in various industrial processes), then significant economic losses would be associated with a high rate of production at present. This by itself is sufficient to justify very high *R/P* ratios.

In conjunction with the preceding series of remarks, chapter 4 argues that the present price of oil, in the light of what we believe today about ultimately recoverable oil supplies, is probably only slightly higher than the economically "correct" price. *If* the world demand for oil is going to outrun the rate of production in only a decade or two, then the price of this commodity must escalate rapidly in order to warn potential consumers not to purchase durable equipment that will be too dependent on oil as a source of energy. Also, if there is going to be a shortage of conventional oil, but no fall in the demand for "oily" substances, then the present price of oil must attain a level that will prompt investments in oil from shale and tar sands, synthetic fuels, and so on.

To look at it from the side of the oil producers, if the governments of certain OPEC countries have drawn the conclusion that their future economic development is to be based on oil products of one type or another, then it behooves them to have a great deal of oil available when the industries for producing these oil products come into existence. One country that seems to be thinking along these lines is Kuwait. Not only are they reducing their production, but also they have informed their customers (and potential customers) that Kuwait intends to refine, domestically, a substantial portion of its oil production by 1984, and anyone interested in buying Kuwaiti oil must also be prepared to pay for refining a part of the purchases in Kuwait. Moreover, that country has intimated that Kuwaiti oil and oil products must be carried on Kuwaiti ships when these are available. By way of emphasizing the position of his government, Kuwait's oil minister, Ali Khalifa Al-Sabah, has made it clear that anyone not interested in these arrangements can take his business elsewhere and that Kuwait reserves for itself the unconditional freedom to adjust prices and quantities to changing conditions on the world oil market.

At this point it is relevant to mention a long talked of scheme for reducing oil imports into the United States that is finally being tried. This involves an import tax on oil. This tax, ostensibly, would reduce the demand for foreign (that is, OPEC) oil, prompt more domestic exploration, and provide some of the money needed to finance alternative energy technologies. All this is well and good, but there is one thing this scheme cannot be expected to do—reduce the price of oil (and thereby OPEC revenues). The more astute of the OPEC directorate have long understood that only a certain amount of purchasing power can be removed from the industrial countries in the form of payments for crude oil. So OPEC's optimal strategy is to obtain these revenues with a minimum sacrifice of oil. Thus the smaller the amount of oil imported by the industrial countries, either the higher the price that will be asked for each barrel or, as in the case cited above, the more oil purchasers will be constrained to use OPEC refining capacity, when such capacity exists. Or, when it does not exist, OPEC clients will be expected to help establish it.

All this leads to one conclusion, the most important of this chapter. Stresses and strains within OPEC, counterstrategies and tactics by the oil-consuming countries, temporary oil surpluses, and brilliant advice from brilliant economists in the industrial countries will lead to a reduction not in oil prices, but in oil production. The only thing that could alter this arrangement would be a revolution in enhancement technology or the discovery of a number of supergiant oil fields that would result in a huge boost in known reserves. In addition, in order to cause a downward pressure on oil prices, these reserves would have to be in the "right" countries and also in oil fields that could be exploited in the near future. Such bonanzas as the recent oil discovery off Newfoundland, the "greatest strike ever," which involves constructing oil platforms in one of the most inhospitable climates in the world as well as digging pipelines below the sea bed in order to keep them from being smashed by icebergs, are of little interest at this stage of the game.

The importance of supergiant oil fields and advanced enhancement techniques also deserves a short clarification. Over half of the world's recoverable crude oil is found in just 33 of the 20,000 to 30,000 oil fields now identified. Of these 33 fields, 17 are classified as supergiant fields, since they contain at least 10 Gbbl of recoverable reserves each. Richard Nehring of the Rand Corporation has estimated that only 4 to 10 supergiant fields, containing 30 to 100 Gbbl remain to be uncovered. But just when this will take place is uncertain, given that it has been about a decade since the last of these Goliaths was confirmed.

Similarly, if we examine the U.S. oil position, we note that of 450 Gbbl of *oil in place*, originally, 115 Gbbl has been produced to date, while only 34 Gbbl of the remaining 335 Gbbl can be classified as exploitable on the basis of present recovery techniques. Thus, at present, the prospect is that two-thirds of the oil that has been discovered in the United States will stay in the ground. In order to lift some of the remaining oil, a great deal of sophisticated scientific work will have to be done on enhancement techniques, or *tertiary recovery processes.* (Primary and secondary processes involve *natural lift* and injecting water or gas into the deposit.) Mainly this calls for lowering the viscosity of the oil by gas injection or the use of chemicals called surfactants, which wash the droplets of oil from the pores of the rocks in which they are fastened. The Royal Dutch Shell Group has published a report in which they claim that tertiary recovery could make another 400 Gbbl of oil available (which is about the amount that has been produced in the world up to now), while some technicians in the Soviet Union are claiming that Soviet oil resources will eventually be increased by a very large amount because of enhancement. At present the United States is producing about 8.5 Mbbl/d of oil, of which 350,000 to 375,000 bbl is being produced with the aid of tertiary recovery methods. Estimates are that by 1985 these figures could be between 500,000 and 2,300,000 bbl/d. To put it another way, if there were a 10 percent improvement in the rate of recovery at present, U.S. oil reserves would double. On the other hand, there are plenty of experts who say that unless there is a major scientific breakthrough, tertiary recovery processes will never be applicable on a large scale.

Chapter 5 presents the reader with a comprehensive review of the concept of elasticities and its application to the demand for energy materials. I present and discuss most of the important short-term forecasts of oil supply and demand. These forecasts are not optimistic where future developments are concerned, and the consensus seems to be that there will not be enough oil available in the first part of the 1980s to support other than a moderate rate of economic growth in the industrial world. I also make extensive reference to a new spectator sport that seems to have broken out among Soviet watchers, both professional and amateur. This concerns the possibility that the Soviet Union will move from being a net exporter to a net importer of oil in the next few years.

The principal theory here is that the Russians have been encountering severe productivity problems in their oil industry, and these are going to get worse. In 1979 Soviet oil production failed to meet the target specified in the previous Five-Year Plan for the third time in three years, and the increase in production in 1979 was the smallest in twenty years. The 1980 oil production target was apparently reduced from 12.8 to 12.1 Mbbl/d, where the 1979 output was 11.7 Mbbl/d. According to the CIA, Soviet production will fall to 10 Mbbl/d or less in 1985, at which time the Soviets could be importing 2 Mbbl/d as compared to 3 Mbbl/d they are exporting now, at least half of which goes to noncommunist countries for hard currencies.

Two issues are relevant here—the state of Russian technology and the actual amount of oil reserves in the Soviet Union. To take the first, such things as Russian exploration and drilling standards are greatly inferior to those of the United States and other noncommunist industrial countries. This is a well-known fact, and even the Soviet government does not bother to deny it. Some people are therefore insisting that because of the present tension between Russia and the noncommunist industrial countries which resulted from the Afghanistan affair, the Soviet Union will be unable to import the technology needed to exploit its more remote oil fields—precisely those fields which must be exploited if a gap is not to eventually open between indigenous demand and supply.

This matter can be dealt with in a few lines. It is now well known that the MIG series of aircraft and such marvels of military technology as the M-24 assault heliocopter were developed without any U.S. or Western assistance. It also seems to be clear that the Russians can construct supersonic aircraft, nuclear power plants, and aircraft carriers. And in 1941 in what can only be called an organizational masterpiece, the Russians relocated about 1,500 industrial plants behind the Urals, moving some of them several thousand miles on a second-rate railway network in the catastrophic opening phase of a major war. That the Soviets are capable of *these* achievements, but cannot ultimately fabricate some bits and pieces of pipe and drills for working in deep reservoirs and cold climates strikes this economist as being absurd, or worse. However, as things stand, this is largely an academic issue. A U.S. firm has built an ultramodern drill-bit plant near the Soviet city of Kuibyshev at a cost of almost $150 million, and apparently that plant needs only a few more machine tools in order to start turning out the high-quality drill bits that have been a bottleneck in the Soviet oil exploration program. Whether the sanctions put on trade with Russia by the Carter administration will prevent the installation of these machine tools remains to be seen, since apparently many U.S. firms have continued to fulfill their contracts with the Russian government. But even if these tools cannot be obtained from the United States, the Russians should experience no trouble obtaining them elsewhere and paying for them with the hard currency earned from their exports of gold and oil. Of course, it has also been said that the sanctions are more apparent than real and that both the U.S. government and the CIA want Soviet oil production to increase so that the Soviets do not compete for the oil of the Gulf.

What about reserves? Oil production in the United States peaked in the early 1970s when the reserve-production ratio was slightly greater than 10, which is considerably below the *R/P* ratio that can be expected in Russia in the next few years. Over a slightly longer time horizon, though, the key issue is just *who* can give an authoritative estimate of the actual amount of oil in the Soviet Union and on its continental shelf, as well as *where* this oil can actually be found. The CIA has continually forwarded claims that Russian reserves of exploitable oil are much smaller than generally anticipated, while a Swedish consulting firm has made quite a name for itself by claiming that Soviet reserves might reach 150 Gbbl, which would put them in the same category as Saudi Arabia. Most observers consider this last estimate exaggerated, claiming that this consulting company has no more competence to pass judgment on Russian energy resources than some of the members of the Swedish Nobel Academy have to teach *or* understand elementary economic theory, which means no competence at all. But at a recent OPEC meeting, one of the most knowledgeable members of the OPEC directorate said that he relies on the Swedes in this matter and that the Russians have plenty of oil.

As far as I can tell, the evidence is mixed; but the most recent Russian oil statistics (which appeared in April 1980) gave no indication whatsoever of an imminent downturn in output. On the contrary, with the exception of Saudi Arabia no country in the world has presented a pattern of such dynamic growth in oil production over the past decade, and the question that must be asked is whether, for geological reasons, a growth in production of this character would have been possible were it not based on very large reserves. In addition, the main trouble with Soviet oil seems to be shortcomings in the infrastructure of western Siberia, which have held up the construction of facilities for transporting oil; an enormous bureaucratic incompetance; and an absence of material incentives for employees in the energy sector. These deficiencies are going to be attacked by significant increases in the remuneration of oil workers and the allocation of more resources to the solution of transportation problems; but most important, it is likely that the internal price of Soviet oil will be allowed to rise to the vicinity of world market prices. Soviet managers buying oil will thus find it in their interests to economize on this product (which they are apparently not doing at present) while oil field managers and technicians will be more inclined to exploit existing installations more intensively.

Some additional background on the Soviet oil situation is provided in the appendix, but we close this discussion by mentioning that there is some talk that a Russian oil shortage will tempt the Soviet leaders to take advantage of the confused situation now existing in the Persian Gulf. This may or may not be so, but recall that the Soviet government refused to become involved in the Greek civil war (just after World War II) because it understood that Greece was vital to the security interests of the West, and under no circumstances would a communist government in that country have been acceptable. I would be very surprised if the present Soviet government did not have the same understanding about the oil of the Middle East. As for the Afghanistan adventure, the presence of the Russian army in Afghanistan might severely reduce the flow of heroin and perhaps a few other mind expanders to Europe and the United States, but basically this has nothing to do with oil.

Some Macroeconomic Problems

Three important macroeconomic topics treated in this book are exchange rates, inflation, and gold. These are linked to oil in the following way. The rising price of oil causes the United States to print more dollars in order to pay for their large imports of that commodity. This can be done because oil is paid for in dollars; and it *is* done because the internal economy of the United States is such at present that the authorities are constrained by political realities to ensure that the country obtains enough energy to keep its motorists happy and that unemployment, which is already high, does not rise further because of an energy shortage. This increased supply of dollars then enters an international economy which already has an excess of that currency and, thus, few private individuals or institutions with a desire to accumulate more. Governments, on the other hand, have less choice: the dollar is *the* international reserve currency, and thus foreign central banks cannot turn it away. They are also under a certain political and moral pressure not to dump this slightly discredited lucre although, insofar as it is possible, many of them are doing so. In a sense, however, they have no choice. By selling dollars they drive down their value, and as a result these countries' oil imports are less expensive. Together with the inflation that is generated by the printing and spreading of dollars, the fall in the value of the dollar has caused a continual slide in the purchasing power of oil. (That is, the amount of goods which can be purchased by a barrel of oil tends to decrease over time.)

After almost seven years of continual oil price rises intermingled with declines in the purchasing power of oil, OPEC has begun to express its dissatisfaction with the way the world economy functions. They are becoming increasingly insistent on maintaining the *real price* (purchasing power) of oil, and in the absence of formal arrangements for bringing this about, they have evidently decided to resort to the simple expedient of adjusting the oil price up more often or, from time to time, reducing production and letting the oil price increase under the pressure of a smaller supply encountering a constant or increasing demand.

This brings us to another rather fascinating quandary. Some major oil producers do not spend all their revenues but recycle them to (or toward) the major oil-consuming countries. Here we need to emphasize one basic fact: The industrial world does *not* need these *petrodollars*, but it does need OPEC's oil; however, much of the oil will not be forthcoming if the financial return (or *yield*) on petrodollars is inadequate. The reader should always keep this in mind, because bankers and sometimes even governments tend to forget it. Take, for instance, the situation on November 14, 1979, when the U.S. government froze all Iranian assets in U.S. banks. Then, on November 21, a banking syndicate led by Chase Manhattan declared Iran in default on a $500 million loan and moved to seize Iranian deposits, since apparently Iran was a week late with an interest payment. Somewhat later, Morgan Guaranty Trust made arrangements to attach Iran's interests in two industrial companies, Krupp and Deutsch Babcock. European bankers immediately went into a panic, claiming that the impounding of Iranian assets set a dangerous precedent and could lead to other large depositors withdrawing their assets and eventually a run on many of the major banks. But it could also lead to something else that is much more serious.

The oil-producing countries with major financial surpluses (which are Saudi Arabia, Kuwait, and the United Arab Emirate) are already, formally or informally, prohibited from purchasing all the factories, skyscrapers, farmlands, and so on in Europe and North America that they might desire to purchase. To a certain extent they are constrained to purchase financial assets, and as a result they certainly are in no position to tolerate assets that are palpably inferior in that they provide an extremely low financial return or are subject to more or less arbitrary confiscation. In such a situation they would have no choice but to solve their portfolio problem by keeping more oil in the ground, which would be extremely bad news for the industrial countries—much worse, in fact, than a few bank failures and a number of near failures where governments, central banks, and international organizations would either come riding to the rescue or smoothe things out *ex post*, as they did with the failures of the Herstatt bank in Germany and the Franklin National Bank of New York.

We now come to a dilemma of the banks that are on the receiving end of the deluge of petro-dollars and other dollars. These banks have to find creditworthy borrowers for this wherewithal, and it so happens that such borrowers do not exist to the extent that they can absorb more than half of these dollars, if that. According to the canons of respectable banking, a creditworthy borrower (in the multimillion-dollar class) would be one who required a loan in order to undertake an economically viable investment, which happens to be one that generates sufficient revenues to cover the cost of borrowing (principal plus interest). Since not enough of these investments are available, tens of billions of dollars are going every year to the governments of LDCs, who use this money to pay for imports of oil and foodstuffs, weaponry, the service of previously acquired debt, and so on. In order to avoid embarrassing silences, the borrowers of this money are not quizzed too closely on the uses to which it will be put; it is usually sufficient to determine that the authorities receiving the loans are sufficiently sophisticated not to transfer a portion of it directly into their own private accounts. At present at least $150 billion is owed by LDCs to private financial institutions, but this sum is rapidly increasing. Some of the industrial countries are also important clients of the international capital markets. Sweden, for example, is developing into one of the largest borrowers, financing not only luxury consumption with loans, but also such quintessentially Swedish nonsense as development aid to the least developable of the LDCs. This concept has been assiduously promoted throughout the last two decades by regiments of unproductive and incompetent bureaucrats, politicians, and academics, many of whom expect to benefit personally from this altruism on the part of the Swedish taxpayers.

Although of late private banks are constantly being warned by their governments to be more restrained in their lending, nobody really expects them to comply. How could they? The Western industrial countries must have oil, and for this to happen, LDCs must borrow money which they cannot use productively or pay back and, as explained in chapter 6 of this book, at interest rates that have been elevated by a high inflation rate. In order to mitigate some of the worst effects of this situation, a great deal of effort is now being expended to work out schemes that will induce more "top flight" credit risks to borrow OPEC money and also to develop techniques for spreading the growing risk involved in international financial transactions. Most of these efforts seem to center on more official lending, and what this involves is the following. Private banks or oil-producing countries lend directly to the governments of the industrial countries, the IMF, or the World Bank—in other words, first-class credit risks. Then these governments or official institutions lend to nonoil LDCs or other near-bankrupt countries. Thus, if anything goes wrong, the problem is foisted onto the taxpayers of the industrial countries (who, in the last analysis, also guarantee the existence of institutions such as the IMF). Although the taxpayers might have some reservations about this arrangement if their

their governments gave them adequate insight into it, it has aroused unrestrained enthusiasm in the board rooms of those financial institutions whose balance sheets reveal an almost touching amount of trust in some of the most visibly corrupt and inept governments in the world.

We can now say a few words about gold. A decade ago the first U.S. Nobel laureate in economics, Paul Samuelson, wrote that gold is of interest only to "French hoarders, Middle East oil sheiks, and underworld figures," while Lord Keynes called gold a "barbarous relic." Viewed from the alpine heights of pure economic theory, both of these contentions are correct; but in terms of that sentimental disease of the mind known as the desire for security, gold is an unmitigated blessing in a world in which even reputable currencies have a tendency to take a nose dive from time to time. It has become common knowledge that anyone with a desire to maintain accumulated wealth should place a portion of her or his assets in gold, or gold bullion coins such as South African Krugerrands, Canadian one-ounce Maple Leafs, or perhaps the bullion coins of various weights that Mexico is in the process of launching. It is also interesting to know that the auric metal, because of its anonymity, has become an important part of the rapidly expanding "black" or subterranean economy where off-the-book trading is the rule. This economy, which is designed to frustrate the tax collector, would probably flourish even if governments were willing, and able, to maintain the value of their currencies. But as it happens, in most countries just now, the cost of keeping the value of a currency stable is an unemployment rate that is politically and, perhaps, morally unacceptable, and in my opinion this will continue to be the case for at least as long as the price of energy is capable of increasing by prodigious amounts every few years.

It is interesting to note that although the flight into gold is often undertaken by people intent on avoiding the ravages of inflation, a mass resort to this type of action will probably accelerate the pace of global inflation. There are a number of reasons for this, and they are enumerated at some length in chapter 6 in this book, but one is particularly interesting for us in the context of the present discussion. As more and more dollars are traded for gold, driving down the value of the dollar, the United States will have to print more of this currency in order to pay its oil bill, because with OPEC in its present mood, they are going to raise the price of oil in phase with the fall in the value of the greenback. But because the dollar is the official reserve asset, this has the effect of raising the supply of money in those countries eventually receiving these dollars (for example, the other industrial countries) unless the governments of these countries employ some particularly sophisticated economic policy, which some of them do, though not all the time. Everything else remaining the same, this increase in money supplies raises prices.

By way of a conclusion I would merely like to assure the reader that any satisfactions or agonies that he or she has experienced over the past seven years in connection with the energy crisis will almost certainly be extended over the rest of this decade. Even so, if all the technical problems faced by humanity were as manageable as those pertaining to energy, this world would indeed be a beautiful place, since the outlook for replacing oil by other, perhaps more satisfactory, energy sources seems very good now, and can only improve. Still, the industrial countries have suffered an enormous and, in some sense, irreparable trauma in the past few years, in that the vulnerability of our civilization has been revealed in a new and unexpected way. Even worse, we may have been provided a glimpse of the material limits of our world, which is an insight that many of us, in principle, would have gladly done without. Since this is no longer possible, a word to the wise (that is, the decision-makers) is perhaps in order. As the international chess grandmaster Savielly Tartakover once remarked "The blunders are all there, just waiting to be made." Even more important, *after* being made, they remain to be made again.

Appendix 8A: A Comment on Oil and the Soviet Union And An American Energy Dilemma

In 1980 the production of crude oil in the Soviet Union should exceed 600 million tons (Mtons) (over 12 Mbbl/d), which will be close to 20 percent of the world output. By comparison, domestic oil production is falling in the United States, where the oil deficit has reached alarming proportions. In terms of reserves, the USSR probably has more than twice the amount of the world's verified reserves as the United States; and the same is apparently true where potentially recoverable reserves are concerned.

This does not mean, however, that the United States is energy-poor. The United States is energy-rich by *any* standard, absolute or relative, and its energy resources are probably far from being fully assessed. But in both energy and natural resources, the Soviet Union is in a class by itself: In resources of all categories, it is potentially, or perhaps already, the richest country in the world; and by the year 2000 Russia will probably display the same economic promise as the United States a century earlier. This assumes, of course, that the bureaucratic incompetence for which such magnanimous tolerance is still shown in that country is alleviated somewhat. Russia possesses the full spectrum of energy wealth: oil, black and brown coal, gas, tar sands, shales, and so on. There is no hard information on uranium deposits, but it is believed that there are large supplies in the Fergana area of central Asia and in northern Siberia and central Kamchatka. On the other hand, the Russians are plentifully endowed with thorium; and thorium-fueled nuclear reactors are certainly on the drawing boards in Russia, and one or more might be on order. In any event, a nuclear reactor using thorium is in operation in the United States, and thorium is considered the world's third most plentiful energy resource after coal and uranium. Finally, although the Soviet geological survey is reputedly very impressive in the matter of non-fuel minerals, oil exploration in the USSR has definitely lagged, and thus there may be some major deposits just waiting to be uncovered.

As indicated in this chapter and earlier in the book, a great deal of controversy surrounds Soviet oil. The (U.S.) CIA claims that Soviet production will fall to 10 Mbbl/d or less by 1985, as a result of exhaustion of fields in the once highly productive Volga/Urals region and the technical and managerial inability of the Soviets to exploit the Siberian deposits. According to the CIA, this will lead the USSR to become a net importer of oil sometime in the fairly near future, thus competing with other nations for

OPEC oil and driving up the price of oil at even a more rapid rate than normally anticipated. It has also been suggested that oil shortages may lead the USSR into political and/or military adventures in or near the Persian Gulf.

The well-known experts on the Soviet economy, Marshall Goldman of Wesleyan University and Robert Campbell of Indiana University, believe the CIA estimates to be too pessimistic, although they cannot quantify their reservations. That the Soviet oil industry is in trouble is well known, but various sectors of the Soviet economy are always in trouble. The point is that time again the Soviets have shown that when they concentrate their efforts, they have a remarkable capacity to break bottlenecks and get results. The quantity and quality of Soviet military hardware should make this abundantly clear. The problem is that Western Sovietologists, being human beings, have a tendency to become emotionally involved with their subject, in one sense or another. "When will the Russians get an atomic bomb?" President Harry Truman, one of the most level-headed and sensible of all the U.S. presidents, once asked physicist Robert Oppenheimer. And when Oppenheimer hesitated, Truman supplied his own answer. "Never!" Plenty of people cherish similar beliefs about the Soviet energy sector and feel that although the Soviets have vast reserves, their system will prevent them from responding in time to this crisis. As far as I am concerned, they will have no problem at all in responding to any energy deficiencies *if* the reserves are actually available. The best technical knowledge and equipment in the world can be and is being purchased by the USSR on the open market, and Moscow is filled with salespeople from the United States and Western Europe who represent organizations prepared to drill for oil and erect factories on the moon if the price is right and who know that thanks to their worldwide sales of gold, oil, and other minerals, the Soviet Union is not short of cash.

If things continue on the present trend, Siberia may eventually be the most important producer of gas and oil in the world. Until recently the Siberian oil industry was concentrated around the huge Samotlor field near the towns of Surgut and Nizhnevartovsk, but production is being extended into medium-sized fields in the same regions, though apparently at considerable expense. The relatively small Tyumen district of western Siberia has been the site of more investment, reckoned in money terms, than any comparable area in the Soviet Union except the city of Moscow. The most important line of development is to the north of Nizhnevartovsk, around an axis provided by the construction of the Surgut-Urengoy railroad. Another thrust is into the Vasyugan swamp of the Tomsk Oblast, although the fields in this region contributed only 8 Mtons of a western Siberian output of 246 Mtons in 1978. Thus, the combined western Siberian output represented 44.4 percent of the total Soviet output in 1978.

The amazing thing about the western Siberian oil industry is the speed at which it has grown—from 15 Mtons in 1970 to 113 Mtons in 1975. About 75 Mtons is being refined in Asian USSR, mostly at the Siberian refineries of Omsk and Angarsk, the eastern refineries of Khabarovsk and Komsomol'sk, and the new refinery at Pavlodar in northern Kazakhstan. A new pipeline has just been constructed to the Belorussian refinery at Novopolotsk, and the intention is to expand pipeline capacity as fast as necessary, since western Siberia is expected to support an increasing share of Soviet oil production. In 1978 the incremental output for western Siberia was 36 Mtons, while the decline in older producing areas was 10 Mtons. The net increase in Soviet production was thus 26 Mtons. In 1979 the western Siberia increment of production was probably several million tons less than in 1978, although the decline in other regions apparently was the same as the previous year. (The exact figures were not available when this was written.) Some people therefore believe that production in this region will be characterized by a decreasing marginal output, while production in areas such as the Volga Urals may decline even more rapidly. If this scenario is valid, then Soviet production may indeed peak toward the middle of the 1980s, and exports decline; but the possibility should not be excluded that if such a peaking did take place, it might be a temporary phenomenon.

A stagnating oil production, followed by declining exports, would be a bitter pill for the Soviet government to have to swallow at present. The economic and political leverage that the Soviets can gain from their exports of energy is, of course, tremendous; and they cannot be expected to give up this advantage unless the geological situation leaves them no choice, which as yet is very uncertain. Eastern Europe is heavily dependent on Soviet oil and gas. For example, the USSR has supplied over 85 percent of the crude oil consumed by Bulgaria, East Germany, and Hungary in 1978; and given the present world situation, these supplies are virtually irreplaceable. Eastern European countries are also investing heavily in the extraction, processing, and transporting of Soviet energy. A number of them have cooperated in the construction of a 1,700-mile, 56-inch-diameter pipeline for bringing natural gas from the Tyumen gas fields in the southern Ural Mountains to the Czechoslovakian border. Increasing amounts of Soviet gas are being used in Eastern Europe, and the Soviet Union sells a considerable amount of energy to Western Europe. Although it is not commonly realized, the Soviet Union is a much more reliable supplier in both the short and the long run than many OPEC countries. Here it should be noted that a kind of revolution may be underway in gas technology in both the Soviet Union and the United States, with the emphasis on deep drilling in methane hydrates (which are layers made of water and gas that solidify naturally underground at certain temperatures and pressures—the Siberian permafrost is thought to be an ideal overlay for gas of this type). If expectations are fulfilled, and among various technicians in both the USSR and

United States there is a great deal of optimism at present, then the world energy picture could be completely altered well before the end of the 1990s.

A final remark on communist economics seems in order here. It has been claimed that the bottleneck in the Soviet sector is an inefficient system of incentives that distorts resource allocation. Physical output is the supreme merit, and there are distinct biases against innovation and quality improvement at all levels of the planning and production process. Certainly the decision that was made a number of years ago not to give factory and project managers more autonomy, particularly in the planning of work-forces and the disposition of revenues, has cost the Soviet economy dearly. But there are also cultural problems that cannot be overlooked. A lack of respect for workmanship and efficiency characterizes almost all aspects of Soviet life except the armed services, the scientific establishment, and perhaps the police and some aspects of the arts. I see no point in making heavy weather of this observation, however, because there are several capitalist countries in almost the same situation, or worse; but unfortunately they lack the enormous natural resource and energy base that may someday give the USSR the "lucky country" designation that is still applicable to Australia.

In 1979 the United States underwent its second oil crisis in less than a decade. Because of the events in Iran, where production declined by 5 million barrels per day (Mbbl/d) of oil between July 1978 and 1979, lines appeared at the gasoline stations once more, and the price of most oil products increased.

In a short but provocative analysis, Verleger (1979) has suggested that both the physical shortages experienced during that period and the price rises were more severe than necessary. In his opinion, the source of the difficulty was the increased stockpiling of refined products which went against the grain of economic rationality; and as a result, some questions must be raised as to the role of the U.S. Department of Energy (DOE) in causing these shortages, since apparently the DOE requested the buildup of energy materials, ordered a regional redistribution of crude oil that might not have been wise, and established gasoline price and allocation regulations that made it profitable to keep supplies from the market. The general drift of Verleger's article is that the DOE was derelict in its duty, but he did not go as far as some critics to suggest that this organization had fulfilled its function, and thus arrangements should be made to bring its activities to a halt.

Now although there is a small probability that the United States will be the victim of an energy catastrophe, such a risk obviously exists; and in the event of a calamity, the domestic economic and political consequences could be momentous. (For instance, they could involve the United States going to war.) As it happens, ordinary markets cannot cope with low-probability–high-loss risks. Nor can they handle certain types of distortions resulting from monopoly power, as exemplified by the OPEC cartel, or the inability of individuals to deal with severe information shortages. The prices formed in the presence of these imperfections are the *wrong* prices for optimal resource allocation to take place. To put it another way, economic efficiency (which involves the allocation of scarce resources in such a way as to achieve the maximum value of total output) requires that the expected costs of an energy shortage be included in the price of energy (and, for that matter, in many other prices). Since there is no practical way to adjust prices accordingly, the energy market is a natural object for policy concern and the ministrations of agencies like the DOE.

People like professor Milton Friedman think otherwise, and unfortunately he does not stand alone. Friedman advocates an immediate closure of the DOE and allowing the price system or the "invisible hand" to take over that department's functions. But as contended above, this might be a mistake. For instance, in 1979 several opinion polls showed that 45 percent of the people in the United States still did not know that the United States was an importer of oil. What this means is that many of these people have purchased, are purchasing, or will be purchasing automobiles and other energy-using durables which might have to stand idle one day because the foreign suppliers of that energy, who are beyond the reach of the U.S. courts, take umbrage with some remark or action of a member of the U.S. government, or some other government with which the United States is associated. *Obviously, no price system can deal with this type of situation;* and so these individuals—and, in fact, more knowledgeable citizens who are in some way dependent on these individuals—occasionally might require someone, or some organization who are in possession of the facts of the energy dilemma and who have the authority to use these facts to promote general equity and efficiency objectives. Establishing such organizations is not only the right but the duty of the government of a civilized country. Beyond a doubt the DOE *has* sponsored some controversial regulations, but whether they deserve the incessant criticism of Dr. Friedman who insisted that the world oil price could not be kept above $10 and even today argues in favor of curing inflation with unemployment while maintaining that the U.S. volunteer army is not a cost/benefit fiasco, remains to be seen.

A Note on the Literature and a Bibliography

Unquestionably the best reading available for anyone interested in energy and energy materials (and natural resources) is to be found in the journals *Energy Policy* and *Resources Policy*, both published by the IPC Science and Technology Press (London). This organization also publishes *Energy Economics*, which is now in its second year and which features both applied and abstract papers. Other important publications dealing mostly with oil are the *Petroleum Economist* and the *Oil and Gas Journal*. For energy matters as such I recommend the *Bulletin of the Atomic Scientists*, which is extremely interested in the nuclear debate, *New Scientist, Science,* and the *Scientific American*. Periodicals that should definitely be regularly examined for easily read, informative, and up-to-date material are *Business Week* and *Fortune*. Some academic-type journals that deal in nontechnical analyses of energy problems are *Intereconomics, Foreign Affairs,* and *Foreign Policy,* with the latter two concentrating largely on some of the political aspects of energy, as well as the *Journal of Energy and Development*. *Energy Communications* is also very useful.

A number of important non-English-language publications should be mentioned here, such as the bulletins of the Centre National de la Recherche Scientifique (Grenoble) and *Energie Wirtschaft* (West Germany). While on this subject I can mention that one of the best libraries for energy references is that of the International Institute for Applied Systems Analysis (Laxenburg-Vienna). A great deal of important work also seems to originate in the vicinity of Cambridge, Massachusetts, with the economics side being handled by Professors Dale Jorgenson and Robert Pindyck; and it is also true just now that the U.S. Department of Energy is sponsoring a great deal of economic research. Much of this research is listed in the "Rapport sur les Recherches Effectuées en 1978 sur L'application des méthodes Mathematiques à l'analyse Economique," which is published by the Economic Commission for Europe. I am also of the opinion that Germany is the country in the industrial world where energy matters receive the most thorough attention. Seminars on energy and energy materials are extremely well attended, and audiences insist on hard answers and not waffling and algebraic demonstrations by academic celebrities making the conference rounds.

The best article that I know on oil is that of Flower (1978). For a background on oil and the world economy, see my book *The International*

Economy: A Modern Approach and the references given there. An interesting examination of the "oil crisis" is presented by Issawi (1978), who also scrutinizes some forecasts of oil supply and demand. A much more detailed scrutiny, however, is that of Ulph (1980).

In the matter of models of cartel behavior, the most realistic construction is clearly that of Pindyck (1978), but an interesting model that has led to many comments and extensions is that of Salant (1976). On the economics of resource extraction, the best elaboration of the basic results is still that of Solow (1974), but the newly published volume *Exhaustible Resources, Optimality, and Trade* by Murray Kemp and N. Long is a very important contribution to this topic. Some forthcoming work of Pindyck treats problems of uncertainty in resource extraction in a way that is fundamentally different from the "cake eating" approach developed by Murray Kemp. Another stimulating and important paper is that of Folie and Ulph (1980), while the best reviews of the various models of oil markets are to be found in the papers of Hammoudeh (1979) and Choucri (1979).

Where energy and its significance are concerned, the reader should examine the work of Percebois listed in the bibliography and chapter 2 of my book *Scarcity, Energy, and Economic Progress.* Possible OPEC strategies are well surveyed by Gateley (1978, 1979) and Doran (1979). The foreign trade of the OPEC countries receives a thorough examination by Gälli (1979) and Lütkenhorst and Minte (1978). The latter economists are interested in an empirical elaboration of the so-called transfer problem, and for a good theoretical background to this important topic see the textbook by Caves and Jones (1977) and the papers of Ron Jones that are listed in the bibliography of this book.

A well-known problem that I have discussed has to do with whether capital and labor are substitutes and complements. The best summary of the basic econometric and theoretical issues, as far as I am concerned, is presented by Berndt and Wood (1977), although a shortened version of this paper has been published in the *American Economic Review* in June 1979; but also see Jorgenson (1978) for a very clear exposition of this and related issues. Jorgenson, who along with Christensen and Lau introduced the translog price functions, has also just coauthored an important working paper on substitution and technical change ("Substitution and Technical Change in Production," by Dale Jorgenson and Barbara M. Fraumeni, Harvard Institute of Economic Research, Discussion Paper 752).

Good general discussions of the "oil crisis" are those of Levy (1978) and Madigan (1979). Although there seems to be some controversy about their conclusions, there is a great deal of valuable information in Stobaugh and Yergin (1979). The best general review of the oil market that I have seen is that of Björk (1978). Special problems associated with the energy price rises are examined by Hu (1978) and Rayment (1980), where Rayment examines certain important issues having to do with energy prices and motoring. A wide range of unrelated but important issues dealing with energy are also taken up by Kern O. Kymn in the papers listed below. His 1978 paper is very important.

For material on the Euromarket see my book *The International Economy: A Modern Approach* and the references cited there. A good article on gold is that of J. Alexander Caldwell (1980). A new and impressive book on natural-resource economics by Lecomber (1979) has recently been published, which includes a discussion of some energy issues; but anyone with a wide interest in energy and natural resources should regard the work of Georgescu-Roegen (1976) as a must. The theoretical background of some of the investment topics treated in this book can be found in the work of Samuelsson, in particular his seminal paper from 1937.

An important article on coal is that of Edward Griffith and Alan Clark in *Scientific American* (1979); and for important information on other energy materials the reader should examine the book of Folie and McColl (1978), which was published by the Centre for Economic Research of the University of New South Wales (Sydney, Australia). Radetzki is completing a book on uranium that treats a number of important economic issues, and Häfele and his colleagues have just completed a book summarizing their work on long-run supply and demand. I should also like to recommend a paper called "U.S. Oil Discovery and Production" by Edward Renshaw and Perry Renshaw (*Futures,* vol. 12, no. 1, February 1980) which makes some very important points about U.S. oil production and its potentialities, and to note that the *OPEC Bulletin* is the source of important and up-to-date news about the world oil economy.

Many topics presented in this book may awaken in the reader a desire to delve deeper into some background materials. An interesting but not too difficult macroeconomic text is the recently published volume of Bronfenbrenner (1979). I still find the intertemporal microeconomic analysis of Alchian and Allen (1964) the best available, and I have attempted to elaborate on a small part of it in my book *Bauxite and Aluminum: An Introduction to the Economics of Non-Fuel Minerals.* Finally, for an overall survey of the topics that I consider to be the most important in economics, I can mention the books of Rostow (1978) and Tuve (1976); and important new work on the macroeconomic effects of North Sea oil has been initiated by Peter Forsyth and John Kay, and on Mexican oil by Richard B. Mancke, in his book *Mexican Oil and Natural Gas* (Praeger, 1979).

Bibliography

Adelman, M.A. *The World Petroleum Market.* Baltimore, Md.: Johns Hopkins University Press, 1972.

______. "The World Oil Cartel: Scarcity, Economics, and Politics." *Quarterly Review of Business and Economics,* Summer 1976.

Alchian, A., and Allen, W. *Exchange and Production Theory in Use.* Belmont, Calif.: Wadsworth, 1964.

Al-Janabi, A. "OPEC Reserves, Production and Exports." *OPEC Weekly Bulletin*, January 1977.

______. "Estimating Energy Demand in the OPEC Countries." *Energy Economics*, April 1979.

Artus, Jacques. "Potential and Actual Output in Industrial Countries." *Finance and Development*, 1979.

Ayoub, Antoine. "Technologie, Matières Premières et Pétrole: Vers un Monopole Bilateral?" *L'Actualitè Economique*, October-December 1977.

Bambrick, Susan. *Australian Minerals and Energy Policy.* Canberra: Australian National University Press, 1979.

Banks, Ferdinand E. *The World Copper Market: An Economic Analysis.* Boston: Ballinger, 1974.

______. *The Economics of Natural Resources.* New York: Plenum, 1976.

______. *Scarcity, Energy, and Economic Progress.* Lexington, Mass.: Lexington Books, D.C. Heath, 1977.

______. *The International Economy: A Modern Approach.* Lexington, Mass.: Lexington Books, D.C. Heath, 1979a.

______. *Bauxite and Aluminum: An Introduction to the Economics of Non-Fuel Minerals.* Lexington, Mass.: Lexington Books, D.C. Heath, 1979b.

Barnett, Donald W. *Minerals and Energy in Australia.* Stanmore, Australia: Cassel, 1979.

Basevi, G. and Steinherr, A. "The 1974 Increase in Oil Prices: Optimum Tariff or Transfer Problem." *Weltwirtschaftliches Archiv*, August 1976.

Bell, Geoffrey. "The OPEC Recycling Problem in Perspective." *Columbia Journal of World Business*, Fall 1976.

Bergman, Lars, and Radetzki, Marian. "How Will the Third World Be Affected by the OECD's Energy Strategies?" *Journal of Energy and Development*, 1979.

Berndt, Ernst R., and Wood, David O. "Engineering and Econometric Approaches to Industrial Energy Conservation and Capital Formation." Resources Paper No. 6, University of British Columbia, December 1977.

Björk, Olle. "Utvecklingen pa den Internationella Oljemarknaden." Department of Economics, University of Stockholm, Skrift Nr 1978:5, 1978.

Bradley, Paul. *The Economics of Crude Petroleum Production.* Amsterdam: North-Holland, 1967.

Bradshaw, Thornton. "My Case for National Planning." *Fortune*, February 1977.

Bronfenbrenner, Martin. *Macroeconomic Alternatives.* Arlington Heights, Ill.: AHM Publishing Corporation, 1979.

Caldwell, J. Alexander. "Gold: The Fundamentals behind the Fury." *Euromoney*, February 1980.

Calvo, Guillermo, and Findlay, Ronald. "On the Optimal Acquisition of Foreign Capital through the Investment of Oil Export Revenues." *Journal of International Economics*, October 1978.

Carman, J. "Comments on the Report Entitled 'The Limits to Growth.'" Address at the University of Brunswick, 1972.

Carlsson, Torsten. "The International Monetary System—Developments and Experience during the Post War Period." *Skandinaviska Enskilda Banken Quarterly Review*, 1979.

Carmoy, Guy de. "Nuclear Energy in France: An Economic Policy Overview." *Energy Economics*, July 1979.

Caves, Richard, and Jones, Ronald. *World Trade and Payments.* Boston: Little, Brown, 1977.

Chapman, Peter. *Fuel's Paradise.* Harmondsworth: Penguin Books, 1975.

Chevalier, J.M. *Det Nye Spill om Oljen.* Trondheim, 1974.

Choucri, Nazli. "Analytical Specifications of the World Oil Market." *Journal of Conflict Resolution*, June 1979.

Clower, Robert. "An Investigation into the Dynamics of Investment." *American Economic Review*, 1954.

Connelly, Philip, and Perlman, Robert. "*The Politics of Scarcity.* London: Oxford University Press, 1975.

Cook, C. Sharp. "Don't Say We Weren't Warned." *Bulletin of the Atomic Scientists*, September 1976.

Cook, Earl. "Limits to the Exploitation of Nonrenewable Resources." *Science*, February 20, 1976.

Cooper, Richard, and Lawrence, Robert Z. "The 1972–73 Commodity Boom." *Brookings Papers on Economic Activity*, No. 3, 1975.

Corden, Warner Max. "Framework for Analysing the Implications of the Rise in Oil Prices." In *The Economics of the Oil Crisis,* edited by T.M. Rybczynski. London: The Macmillan Press, 1976.

Dasgupta, Biplab. "Oil Prices, OPEC, and the Poor Oil Consuming Countries." In *Future Resources and World Development*, edited by Paul Rogers. New York: Plenum, 1976.

Davidson, Paul. "Fiscal Policy Problems of the Domestic Crude Oil Industry." *American Economic Review*, March 1963.

Doran, Charles F., and Hopkins, John. "Three Models of OPEC Leadership and Policy in the Aftermath of Iran." *Journal of Policy Modeling*, 1979.

DuMoulin H., and Eyre, J. "Energy Scenarios: A Learning Process." *Energy Economics*, April 1979.

Dunham, Kingsley. "How Long Will Our Minerals Last?" *New Scientist*, January 1974.

Erikson, Edward W. and Grennes, Thomas J. "Arms, Oil, and the American Dollar." *Current History,* May/June 1979.
Fisher, Anthony C. "On Measures of Natural Resource Scarcity." International Institute for Applied Systems Analysis, February 1977.
Fisher, Franklin M. *Supply and Costs in the U.S. Petroleum Industry.* Washington: Resources for the Future, 1964.
Fisher, John. *Energy Crisis in Perspective.* New York: John Wiley & Sons, 1974.
Flower, Andrew R. "World Oil Production." *Scientific American,* March 1978.
Folie, Michael, and McColl, Gregory. *The International Energy Situation Five Years after the OPEC Price Rises.* Sydney: Centre for Economic Research, 1978.
______, and Ulph, Alastair M. "Outline of an Energy Model for Australia." CRES Working Paper, 1976.
______, and Ulph, Alastair M. "The Use of Simulation to Analyze Exhaustible Resource Cartels." Canberra: Australian National University, 1978.
______, and Ulph, Alastair M. "Energy Policy for Australia." in R. Webb and R. Allen, *Australian Industrial Policy.* Allen and Unwin, 1980.
Forester, Jay W. *World Dynamics.* Cambridge, Mass.: Wright-Allen Press, 1971.
Fritsch, Bruno. "The Zencap-Project: Future Capital Requirements of Alternative Energy Strategies Global Perspectives." Fifth World Congress of the International Economic Association, Tokyo, 29 August–3 September 1977.
Gälli, Anton. "The Foreign Trade of the OPEC States." *Intereconomics,* November/December 1979.
Garnault, Ross, and Clunies-Ross, A. "Uncertainty, Risk Aversion, and the Taxing of Natural Resource Projects." *Economic Journal,* June 1975.
Gately, Dermot. "The Possibility of Major Abrupt Increases in World Oil Prices by 1990." Discussion Paper Series, New York University, Faculty of Arts and Sciences, Department of Economics, May 1978.
______. "OPEC Pricing and Output Decisions." paper prepared for the Conference on Applied Game Theory, Institute for Advanced Studies, Vienna, June 13–16, 1978.
______. "The Prospects for OPEC Five Years after 1973/74." *European Economic Review,* no. 12, 1979.
Georgescu-Roegen, Nicholas. *The Entropy Law and the Economic Process.* Cambridge, Mass.: Harvard University, 1971.
______. *Energy and Economic Myths.* New York: Pergamon, 1976.
Griffith, Edward D., and Clarke, Allan W. "World Coal Production." *Scientific American,* January 1979.
Haefele, Wolf. "Global Perspectives and Options for Long-Range Energy

Strategies." Keynote address at the Conference on Energy Alternatives, East-West Center, Honolulu, Hawaii, January 9-12, 1979.

Hammoudeh, S. "The Future Price Behavior of OPEC and Saudi Arabia: A Survey of Optimization Models." *Energy Economics*, July 1979.

Harlinger, Hildegard. "Neue Modelle für die Zukunft der Menscheit." IFO-Institut für Wirtschaftsforschung, Munich, February 1975.

Hayes, Earl T. "Energy Resources Available to the United States, 1985 to 2000." *Science*, January 1979.

Herin, Jan, and Wijkman, Per Magnus. "Den Internationella Bakgrunden." Stockholm: Institut för Internationella Ekonomi, 1976.

Hippel, Frank von, and Williams, Robert. "Solar Technologies." *Bulletin of the Atomic Scientists*, November 1975.

Hoffmeyer, Martin, and Neu, Axel. "Zu den Entwicklungsaussichten der Energiemärkte." *Die Wirtschaft*, Heft 1, 1979.

Hotelling, Harold. "The Economics of Exhaustible Resources." *Journal of Political Economy,* April 1931.

Hu, Shei-Yuan David. "The Copper Commodity Model and Energy Issues." Unpublished dissertation, University of Pennsylvania, 1978.

Issawi, Charles. "The 1973 Oil Crisis and after." *Journal of Post Keynesian Economics,* Winter 1978/79.

Janssen, E.R. "Le Prix du pétrole Brut et des Produits Pétroliers et leur Evolution en Europe." *Recherchers Economiques de Louvain*, March 1978.

Jones, Ronald W. "Presumption and the Transfer Problem." *Journal of International Economics,* August 1975.

______. "Terms of Trade and Transfers: The Relevance of the Literature." In *The International Monetary Sytem and the Developing Nations.* AID, U.S. Department of State, 1976.

______. "*International Trade: Essays in Theory.*" Amsterdam: North-Holland, 1979.

Jorgenson, Dale W. "The Role of Energy in the United States Economy." *National Tax Journal*, September 1978.

______., and Hudson, Edward. "Economic Analysis of Alternative Energy Growth Patterns, 1975-2000." In *A Time to Choose*, edited by D. Freeman, et al. Cambridge, Mass.: Ballinger, 1974.

Kapitza, Peter. "Physics and the Energy Problem." *New Scientist*, October 1976.

Kellog, Herbert. "Sizing Up the Energy Requirements for Producing Primary Metals." *Engineering and Mining Journal*, April 1977.

Kemp, A.G., and Crichton, D. "North Sea Oil Taxation in Norway." *Energy Economics*, October 1979.

Kymn, Kern O. "Interindustry Energy Demand and Aggregation of Input-Output Tables." *The Review of Economics and Statistics, August 1977.*

______. "A Contribution toward a Correct Understanding of the Welfare Analysis of Foreign Supply Interruptions." *Energy Communications*, vol. 4, 1978.

______, and Page, Walter P. "Some Interindustry Empirical Evidence on the Question of Stockpiling Crude Oil." *Energy: The International Journal*, October 1978.

______, and Page, Walter P. "U.S. Interindustry Cost Structure of the Energy Sectors." *Energy: The International Journal*, Spring 1979.

Lapp, Ralph E. "We May Find Ourselves Short of Uranium Too." *Fortune*, October 1975.

Lave, L.B., and Seskin, E.P. "Air Pollution and Human Health." *Science*, August 1970.

______. "Health and Air Pollution." *Swedish Journal of Economics* March 1971.

______. "Acute Relationships among Daily Mortality, Air Pollution, and Climate." In *Economic Analysis of Environmental Problems*, edited by E.S. Mills. 1975.

Lecomber, Richard. *The Economics of Natural Resources*. London: The MacMillan Press, 1979.

Leontief, W. *The Future World Economy*. New York: United Nations, 1977.

Levy, Haim, and Sarnat, Marshall. "The World Oil Crisis: A Portfolio Interpretation." *Economic Inquiry*, September 1975.

Levy, Walter J. "The Years that the Locust Hath Eaten: Oil Policy and OPEC Development Prospects." *Foreign Affairs*, Winter 1978.

Lovins, Amory. "Energy Strategy: The Road Not Taken." *Foreign Affairs*, October 1976.

Lütkenhorst, Wilfried, and Minte, Horst. "Probleme des Monetären und Realen Transfers." In *Die Energiekrise: Fünf Jahre Danach*, edited by Manfred Tietzel. Bonn: Neue Gesellschaft GmbH, 1978.

______. "The Petrodollars and the World Economy." *Intereconomics*, March/April 1979.

MacKay, G.A. "Uranium Mining in Australia." In *International Resource Management*, edited by J.T. Woodcock. Canberra: Austrailian Institute of Mining and Metallurgy, 1978.

Madigan, Alan L. "Oil Is Still Cheap." *Foreign Policy*, Summer 1979.

Manne, Alan, Richels, Richard, and Weyant, John P. "Energy Policy Modeling: A Survey." *Operations Research*, January-Febraury 1979.

Mayorga-Alba, Eleodoro. "L'Exploration-Production dans la Nouvelle Conjoncture Pétrolière: Fiscalitès et Rentabilitès." *Recherches Economiques de Louvain, March 1978.*

McKern, R.B. Multinational Enterprises and National Resources. Sydney: McGraw-Hill, 1976.

McKie, James W. "Market Structure and Uncertainty in Oil and Gas Exploration." *Quarterly Journal of Economics*, November 1960.

Moran, Theodore. "Why Oil Prices Go Up: OPEC Wants Them to." *Foreign Policy*, Winter 1976–1977.

Murcier, Alain. "Carter Face aux Petroliers." *L'Expansion*, December 1979.

Nordhaus, W.D. "Resources as a Constraint on Growth." *American Economic Review*, May 1974.

______. "The Allocation of Energy Resources." *Brookings Institution Papers*, 1973.

Nulty, Peter. "When We'll Start Running Out of Oil" *Fortune*, October 1977.

Page, Walter P., and Kymn, Kern O. "Cartel Policy and the World Price of Oil: An Explanation?" *Energy Communications*, 1978.

Pearce, David. *Environmental Economics*. London: Longmans, 1976.

Percebois, Jacques. "Energie, Croissance et Calcul Economique." *Revue Economique*, May 1978.

______. "A Propos de queloques Concepts Utilisés en Economie de l'Energie" In *Economies et Sociétés*, no. 3, July–September 1978.

______ "Is the Concept of Energy Intensity Meaningful?" *Energy Economics*, July 1979.

Petersen, Ulrich, and Maxwell, Steven R. "Historical Mineral Production and Price Trends." *Mining Engineering*, January 1979.

Pindyck, Robert. "OPEC's Threat to the West." *Foreign Policy*, Spring 1978.

______. "Gains to Producers from the Cartelization of Natural Resources." *The Review of Economics and Statistics*, May 1978.

Radetzki, Marian. "Will the Long Run Global Supply of Industrial Minerals Be Adequate?" Paper presented at the Fifth Congress of the International Economic Association, Tokyo, September 1977.

____. *The International Uranium Market in the 1970s: A Study of Violent and Unpredictable Commodity Price Swings and Their Causes*. Stockholm: Institute for International Economic Studies, 1979.

______. and Zorn, Stephan. *Financing Mining Projects in Developing Countries*. London: Mining Journal Books, 1979.

Rayment, Paul. "Petrol Prices, Conservation, and Macro-Economic Policy." Economic Commission for Europe, February 1980.

Renton, Anthony. "A Bigger Bonanza." *New Scientist*, September 23, 1976.

Ross, Marc H., and Williams, Robert H. "Energy Efficiency: A Most Underrated Energy Resource." *Bulletin of the Atomic Scientists*, November 1976.

Rostow, W.W. *Getting From Here to There: A Policy for the Post-Keynesian Age*. New York: Macmillan, 1978.

Russell, Robert W. "Governing the World's Money: Don't Just Do Something, Stand There." *International Organization*, Winter 1977.

Rustow, D.A., and Mugno, John. *OPEC: Success and Prospects*. New York: New York University Press, 1976.

Salant, Stephen W. "Exhaustible Resources and Industrial Structure: A Nash-Courant Approach to the World Oil Market." *Journal of Political Economy*, 1976.

Samuelsson, Paul Anthony. "Some Aspects of the Pure Theory of Capital." *Quarterly Journal of Economics*, May 1937.

Siebert, Horst. "Ershöpbare Ressourcen." *Wirtschaftsdienst*, no. 10, 1979.

Slesser, Malcolm. *Energy in the Economy*. London: Macmillan, 1978.

Solomon, Robert. *The International Monetary System, 1945*–1976. New York: Macmillan, 1977.

Solow, John. "A General Equilibrium Approach to Aggregate Capital-Energy Complementarity." *Economic Letters*, no. 2, 1979.

Solow, Robert. "Richard T. Ely Lecture: The Economics of Resources or the Resources of Economics." *American Economic Review*, May 1974.

______. "Resources and Economic Growth." *The American Economist*, Fall 1978.

Spengler, Joseph. "Population and World Hunger." *Rivista Internazionale di Scienze Economiche E. Commerciali*. December 1976.

Steen, Peter. *Om Oljeförsörjningen*, Stockholm: Sekretariatet för Framtidsstudier, 1977.

Stobaugh, Robert, and Yergin, Daniel. "After the Second Shock: Pragmatic Energy Strategies." *Foreign Affairs,* Spring 1979.

Tuve, George L. *Energy, Environment, Population, and Food: Our Four Interdependent Crises*. New York: Wiley-Interscience, 1976.

Ulph, Alistair M. "World Energy Models—A Survey and Critique." *Energy Economics*, January 1980.

______, and Folie, Michael. "Gains and Losses to Producers from Cartelisation of and Exhaustible Resource." Canberra: CRES Working Paper, 1978.

Van Duyne, Carl. "Commodity Cartels and the Theory of Derived Demand. *Kyklos*, 1975.

Vann, Anthony, and Rogers, Paul. *Human Ecology and World Development*. New York: Plenum, 1974.

Verleger, Philip K. "The U.S. Petroleum Crisis of 1979". *Brookings Papers*, no. 2, 1979.

Walters, Alan Rufus. "The Economic Reason for International Commodity Agreements." *Kyklos*, 1974.

Weintraub, E. Roy. *Microfoundations*. Cambridge, England: Cambridge University Press, 1979.

White, Norman A. "The International Availability of Energy Minerals." *CIM Bulletin*, September 1978.

Index

About the Author

Ferdinand E. Banks is associate professor at the University of Uppsala and visiting professor at the Centre for Policy Studies, Monash University, Melbourne. During 1978 he was Professorial Fellow in economic policy at the Reserve Bank of Australia and visiting professor in the Department of Econometrics, the University of New South Wales. Professor Banks attended Illinois Institute of Technology and Roosevelt University and received the B.A. in economics. After serving with the U.S. Army in the Orient and Europe, he worked as an engineer and systems and procedures analyst. He received the M.Sc. and Fil. Lic. from the University of Stockholm and the Fil. Dr. from the University of Uppsala. He taught for five years at the University of Stockholm; was senior lecturer in economics and statistics at the United Nations African Institute for Economic and Development Planning, Dakar, Senegal; and has been consultant-lecturer in macroeconomics for the OECD in Lisbon. From 1968 to 1971 Professor Banks was an econometrician for the United Nations Commission on Trade and Development in Geneva; and he has also been a consultant on planning models and the steel industry for the United Nations Industrial Organization in Vienna. His previous books are *The World Copper Market: An Economic Analysis* (1974); *The Economics of Natural Resources* (1976); *Scarcity, Energy, and Economic Progress* (1977); *The International Economy: A Modern Approach* (1979); and *Bauxite and Aluminum: An Introduction to the Economics of Nonfuel Minerals* (1979). He has also published forty-five articles and notes in various journals and collections of essays, as well as a large number of reviews.

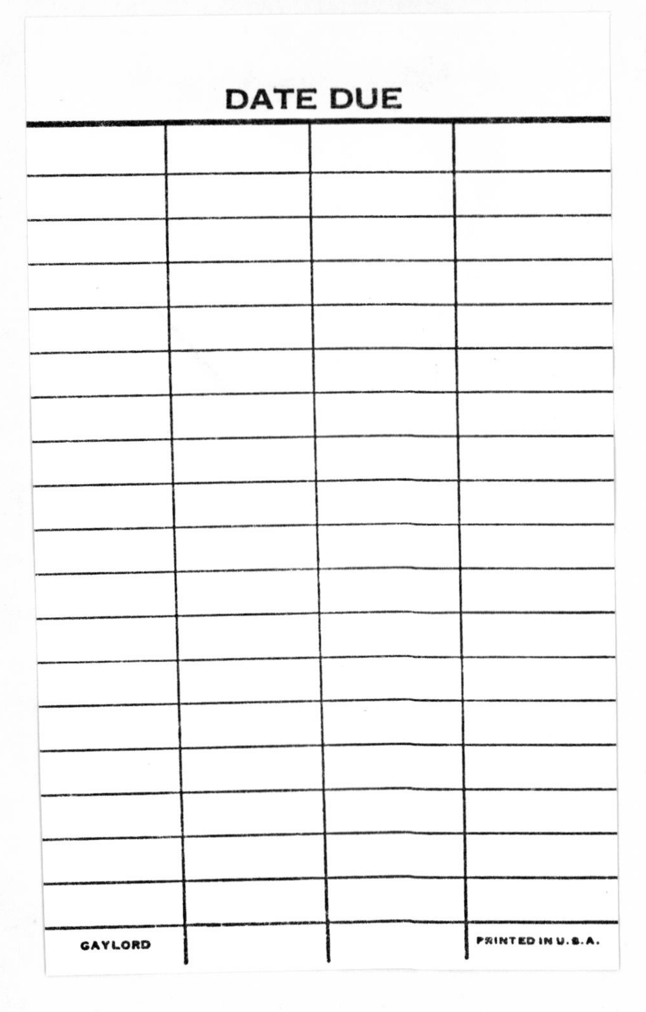
DATE DUE
GAYLORD
PRINTED IN U.S.A.